TECHNICAL KNOWLEDGE IN EARLY MODERN JAPAN

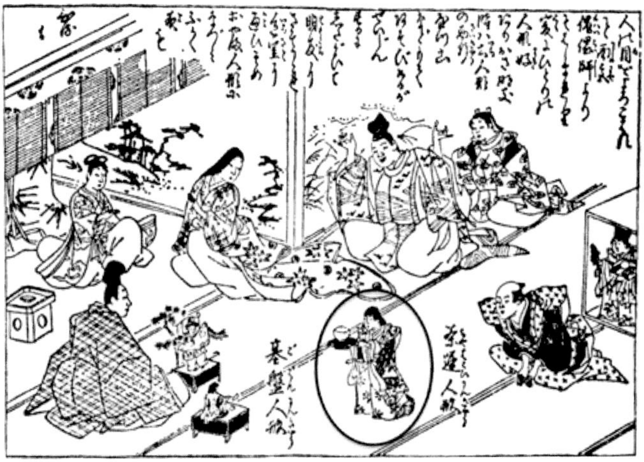

The 'Tea-serving boy' – See Chapter 2

Technical Knowledge in Early Modern Japan

Edited by

Erich Pauer
University of Marburg

&

Ruselle Meade
Cardiff University

Routledge
Taylor & Francis Group
LONDON AND NEW YORK

TECHNICAL KNOWLEDGE IN EARLY MODERN JAPAN

First published in 2020 by RENAISSANCE BOOKS

Published 2025 by Routledge
4 Park Square, Milton Park, Abingdon, Oxon OX14 4RN
605 Third Avenue, New York, NY 10158

Routledge is an imprint of the Taylor & Francis Group, an informa business

British Library Cataloguing in Publication Data
A catalogue record for this book is available from the British Library

ISBN: 9781912961009 (hbk)
ISBN: 9781041187042 (pbk)
ISBN: 9781003704942 (ebk)

Cover illustration
Smelting furnaces in the Sado gold and silver mines, early modern period. (Source: Picture scroll Sado kōzan kingin saisei zenzu jō. Courtesy of the Mining History Document Center (montan.dok) at the German Mining Museum Bochum (DBM) 030330914000)

For Product Safety Concerns and Information please contact our EU representative: GPSR@taylorandfrancis.com
Taylor & Francis Verlag GmbH, Kaufingerstraße 24, 80331 München, Germany

Contents

Acknowledgments

THIS VOLUME BRINGS together some of the lectures presented at a conference entitled 'The generation and dissemination of technical knowledge in Japan from the Edo to the Meiji Period', held from 18 to 20 September 2018 and organised by the Centre Européen d'Études Japonaises d'Alsace (CEEJA) in Kientzheim, Alsace, France.

In the expectation that these lectures will be most useful to readers who do not have access to scholarship in Japanese, several changes and additions have been made for readers unfamiliar with the history of Japan, especially concerning economic history and the history of technology in that country. An outline of the evolution of the history of technology in Japan has also been added.

The conference was made possible by the support of several institutions in Japan and France. The editors would especially like to thank the Toshiba International Foundation, which generously supported the conference. Our thanks also go to the Département Haut-Rhin and to CEEJA, which helped in various ways to make the conference a reality.

Paul Norbury, publisher of Renaissance Books, has kindly agreed to include a large number of illustrations in this book to support the understanding of the sometimes very technical explanations. He also advised on the publication process and helped overcome obstacles in preparing this book for publcation.

Erich Pauer
Ruselle Meade

Introduction

Remarks on the History of Technology in Japan

Erich Pauer

1. HISTORY OF TECHNOLOGY: A LARGELY UNRESEARCHED FIELD IN JAPANESE STUDIES?

THE HISTORY OF technology is still a largely unresearched field in Japanese studies outside Japan. This Introduction, therefore, begins with some general remarks on the development of the history of technology in Japan.

Do an internet search for 'technology' or 'history of technology' together with 'East Asia', and up pop lots of entries on China, with considerably fewer on Japan, and with Korea trailing along behind. But such numbers tell only half the story. The history of science, and to a lesser extent that of technology, have strong support in countries in the region, and are characterised by lively publishing activity there.[1] Outside these countries, however, China holds a pre-eminent position in terms of research on the history of science and technology, at university institutions and above all in publications. This can be traced back – among other reasons – to the major influence of the comprehensive works of Joseph Needham (1900–1995). His multi-volume series on *Science & Civilisation in China*, published from 1954 and continued up to the present, set a high standard early on and influenced numerous sinologists.

1 In China, Japan and Korea, the 'history of technology' does not exist as a separate field. Instead, one usually speaks of the 'history of science and technology', or *kagaku gijutsu-shi* in Japanese. Although it appears that both fields of research are equally important, the history of technology is de facto mostly treated as a subfield of the history of science.

Nothing comparable has ever been produced for Japan. Although studies of the history of technology in Japan also go back a long way[2] and are represented in the academic field through institutions, publications and journals, scholars dealing with Japan have not succeeded in establishing as strong a tradition as those dealing with China.

2. PRE-WAR: EVOLUTION OF THE HISTORY OF TECHNOLOGY IN JAPAN

The beginnings of the history of Japanese technology are fed by various currents, starting with early studies of Japan's industrial development in the Meiji period. Most of them were officially initiated descriptions or documentations of industrial-technical changes in individual enterprises (e.g., naval shipyards, military arsenals), entire sectors, such as communications, or other public institutions. As a rule, such works were neither written by trained historians of technology – such an academic discipline was not established in Japan until after the Second World War – nor were they conceived as scientific or technical-historical works at all. The aspect of the industrial boom and questions of further development in the future were brought to the fore in these works, and this implied a focus on technology, too. However, the treatment of what was recorded as 'technical objects' and 'technical processes' in the broadest sense of the term, in an enterprise or production process, makes these works important today as sources for the history of technology. This also applies to the work *Nihon kōgyō-shi* (Japanese industrial history) by Yokoi Tokifuyu (1860–1906), which was published in 1887 and repeatedly reprinted until the 1930s. But in this case the author, a trained historian, who did not conceive his work as an official documentation, freed industrial history from the fetters of the official tradition of historiography that paid homage to description, as practised for example at Tōkyō Imperial University.[3]

Japan's transition from an agrarian to an industrial state around 1920 stimulated a new interest in the country's technical development and its roots. Especially in the 1920s, a stronger emphasis on

---

2 A first series on Japanese technological history appeared as early as 1914. Written by civil engineers, it was intended to illustrate the achievements of civil engineers in the past and present to a wider public.

3 Yamazaki Toshirō, 'Japan's Contribution to the Modern History of Technology', in *Japanese Studies in the History of Science* 1, pp. 45–47, esp. p. 45.

the historical dimension became noticeable in numerous areas. It is therefore not surprising that it was in this decade – not least under the impression of having caught up with Western industrial countries – that attempts were made to portray the successes achieved in the industrial sector during the Meiji period, which was already considered the formative period of modern industrial Japan. In a ten-volume series *Meiji kōgyō-shi* (History of Industry in the Meiji Period),[4] all the major modern industries were covered. Although this series dealt with industrial history, the editor – an academic society – made sure that works by experts on science and technology from the respective industries were included. Other academic societies and individuals followed this example and published studies on the development of individual branches during the Meiji period.

A different strand can be discerned in a study on the traditional *Tatara* iron-smelting method, based on observations and records which Tawara Kuniichi (1872–1958), a trained metallurgist, had made in 1898 and 1899 in the Chūgoku region. Published in 1933 (including the transcript of an original source from the Edo period), this book marks, as it were, the transition from works oriented towards the description of industrial success to clearly historical studies, which at the same time took technical aspects more into account.[5]

Parallel to this kind of industrial history, another trend developed in the realm of philosophy, which proved to be very important for the development of studies on the history of technology: the interest in materialism (*yuibutsu-shugi*) and later increasingly also in the historical materialism of Marxian influence. The crises of the 1920s, culminating in the world economic crisis of 1929–1932, were widely blamed on capitalism and its contradictions, and triggered a growing interest in Marxist thought and corresponding literature. While socialist and Marxist thought were soon increasingly suppressed by the government, discussions about the concept of technology or technique (and thus implicitly about the 'productive forces' in the Marxist sense) could be carried on and reached a first peak in the early 1930s.

The debate on the concept of technology that began in 1932 on a philosophical level was mostly led by members of the 'Study Group on Materialism' (Yuibutsu-shugi kenkyū-kai), which was founded in 1932 by the neo-Kantian philosopher Tosaka

4 Published between 1925 and 1930 by Kōgakkai, Tōkyō.

5 Tawara Kuniichi, *Korai no satetsu seiren-hō (Tatara-buki seitetsu-hō)* ('Traditional iron-smelting (The tatara-buki smelting technique')), Tōkyō: Maruzen 1933.

Jun (1900–1945) together with the philosopher of science and technology Saigusa Hiroto (1892–1963), the historian of science and technology Aikawa Haruki (1909–1953), and the historian of science Oka Kunio (1890–1971). They all published contributions about technology and technique in the journal *Yuibutsu-ron kenkyū* (Studies of materialism) that appeared from 1932 to 1938. This is a major source for the reconstruction of the different views on the subject. The debate, conducted at first mostly along the lines of abstract Western discourse and unrelated to the current or past situation in Japan, led to two different definitions: Tosaka Jun granted technology an objective (*busshitsu-teki gijutsu*; material objects, processes) and a subjective (*kannen-teki gijutsu*; spiritual, human/social) dimension, but found few supporters. The view of technology as an organization of tools and work equipment, as advocated by Aikawa Haruki, ultimately won the approval of those circles interested in technical development – reduced to tools, machines and processes – and made its mark on the following discussions.

In the intensive debates on technology, the role of the history of 'technique' or 'technology' in the history of humankind became more and more a central point, a central question. In 1935 Tosaka and Oka published a bibliography on technology in *Yuibutsu-ron kenkū* (no. 28), which for the first time included a section on the history of technology; however, this did not mention a single Japanese scholar. Other milestones were two publications by Oka in 1937. An article on *Gijutsu no rekishi-teki kenkyū* (The historical study of technology), appeared in the leading intellectual journal *Chūō kōron*,[6] which made the term 'history of technology' known to a wider audience. The other article, *Gijutsu-shi ni tsuite – gijutsu-shi jōron hō-i* (On a history of technology – Supplement to the introduction to the history of technology),[7] published in the journal *Yuibutsu-ron kenkyū*, specified his reflections on the history of technology and can be regarded as a programmatic paper for this field. However, before such efforts could bear fruit, the journal *Yuibutsu-ron kenkyū* closed in February 1938 for political reasons, and some of its leading authors, such as Tosaka Jun, faced reprisals.

6 Oka Kunio, 'Gijutsu no rekishi-teki kenkyu- – Gijutsu-shi-ron josetsu' (The historical study of technology – Introduction to the theory of the history of technology), in *Chūō kōron* no. 7 (no. 596), 1937, pp. 52–61.

7 Oka Kunio, 'Gijutsu-shi ni tsuite – Gijutsu-sho joron hōi – jō/ge' (On the history of technology – An amendment to the introduction to the history of technology – parts one and two), in *Yuibutsu-ron kenkyu-* no. 57, 1937, pp. 19–26 and no. 58, 1937, pp. 142–148.

While between 1935 and 1937 the philosophical theorists of the 'Study Group on Materialism' continued to work on a clarification and delimitation of the concept of technology and the history of technology, articles by other authors appeared in *Yuibutsu-ron kenkyū*, and other journals, which can be regarded as early contributions to a classical history of technology. They often had a focus on Japan and dealt with topics like the machine tool industry, or electrical engineering.[8]

The number of publications on the history of science and technology increased rapidly in the second half of the 1930s. In addition to translations of foreign writings on the history of technology, more popular educational scientific treatises were also published in order to increase public acceptance of science and technology and a rational (resource-saving) production and lifestyle based on them. The journal of this movement, *Kagaku-shugi kōgyō* (Scientific industry), was founded by Ōkōchi Masatoshi (1878–1952), an engineer and director of the Riken research institute.[9] Saigusa Hiroto, among others, also published articles in it, although it increasingly fell into nationalist waters.

A bibliography compiled by Aikawa Haruki in 1940, in which technology was classified under three categories, philosophy, economics and natural science, and the history of technology, was assigned to the respective fields, lists more than 100 technology-related publications, journal articles and books by about 65 authors, including Japanese authors, for the period between 1935 and 1940.[10] This bibliography and another list with titles concerning the history of science and technology, which reaches

---

[8] Okumura Shōji (pseud. Ishikawa Yoshikazu), 'Dōgu kara kōsaku kikai e no hattatsu' (The development from tools to machine tools), in *Yuibutsu-ron kenkyū*, no. 38, 1935, pp. 106–14; Oka Kunio, 'Waga kuni ni okeru denki gijutsu no hattatsu' (The development of electrical engineering in Japan), in *Yuibutsu-ron kenkyū*, no. 52, 1937, pp. 110–132 (Part I), no. 53 1937, pp. 132–149 (Part II), no. 55, 1937, pp. 106–22 (Part III), no. 56, 1937, pp. 114–134 (Part IV); Satō Kazuo, 'Nihon ni okeru kōsaku kikai no saikin no hattatsu' (On the recent development of machine tools in Japan), in *Yuibutsu-ron kenkyū*, no. 57, 1937, pp. 77–93; Chūjō Yuriko, 'Rajio ogon jidai' no teichō' (The weakness of the 'Radio's golden age'), in *Yuibutsu-ron kenkyū*, no. 58, 1937, pp. 34–40; Tsuruta Michio, 'Meiji-jidai ni okeru waga kuni kōsaku kikai kōgyō hatten no tokushitsu' (The peculiarity of the development of the machine tool industry in Japan in the Meiji era), in *Gakugei*, no. 67, 1938, pp.124–133.

[9] Rikagaku kenkyūjo (Institute of Physical and Chemical Research), is a scientific research institute, founded in 1917.

[10] Aikawa Haruki, *Gendai gijutsu-ron* (On contemporary technology), Tōkyō: Mikasa shobō,1940, pp. 293–305.

beyond the period covered by Aikawa,[11] include the following publications, which formed the core of the newly developing scientific history of technology in Japan:

1938 Masumoto Setsu, *Gijutsu-shi* (History of technology)
1940 Saigusa Hiroto, *Gijutsu-shi* (History of technology)
Yamamoto Saburō, *Nihon kindai gijutsu-shi* (Modern history of technology in Japan)
1942 Yazaki Dan, *Gijutsu bunka-shi* (Cultural history of technology)
Aikawa Haruki, *Gijutsu-ron nyūmon* (Introduction to the theory of technology)
Tamura Eitarō, *Nihon no gijutsu-sha* (Japanese engineers)
Minamitane Yasuhiro, *Nihon no kōgyō-shi* (Industrial history of Japan)
Kamo Giichi, *Gijutsu-hattatsu-shi* (History of the development of technology)
Okubo Tsunetsugu, *Nihon kagaku gijutsu-shi-wa* (Talks on Japanese science and technology
Tamura Eitarō, *Nihon denki gijutsu-sha den* (Biographies of Japanese electrical engineers)

Not all of these books deal with Japan. For example, Masumoto Setsu's *Gijutsu-shi*, mentioned first in the list, describes exclusively the development of technology in the West. So, the first notable history of technology concerning Japan is Saigusa Hiroto's book *Gijutsu-shi* (History of technology), a pioneering achievement published in 1940.[12] Saigusa reveals the basis of his understanding of technology in a theoretical discourse. He refers to the 'processual' nature of technology, whereby technology is understood not only as a means of working, but also as a means of achieving something.[13] The history of technology in this sense cannot be reduced to the history of the tool or the machine. This is what sets Saigusa apart from other contemporaries who strictly pursued the concept of technology as a working tool.

His book comprises a foreword, two parts (with eight and ten chapters respectively), an appendix and a bibliography. The

[11] Kikuchi Toshiyoshi, 'Nihon kagaku-shi gakkai no sōritsu' (The foundation of the Japanese Society for the History of Science), in *Kagaku-shi kenkyū* (Journal of history of science, Japan), vol. 32(187), 1993, pp. 166–173, esp. pp. 166–167.

[12] Saigusa Hiroto, *Gijutsu-shi* (History of technology) (Gendai Nihon bunmei-shi, vol. 14), Tōkyō: Tōyō keizai shinpō-sha, 1940. Reprint: Saigusa Hiroto, 'Gijutsu-shi' (History of technology), in *Saigusa Hiroto chōsaku-shū* (Collected works of Saigusa Hiroto), vol. 10, pp. 13–288, 1973, Tōkyō: Chūō kōron-sha.

[13] Saigusa, *Gijutsu-shi* (History of technology), 1940, p. 7.

structure and content of this first major work on the history of technology reveal a number of special features. For example, Saigusa does not, like most of the Japanese economic historians of the 1930s, resort to the stage scheme of historical materialism, despite the materialistic background that he has made clear. Nor does he link the development of technology in Japan to the scheme of the English industrial revolution, which often served as a model for the treatment of Japan's industrial revolution and thus led to the (widespread, though false) view of Japan jumping from the middle ages into modern times. Moreover, the textile industry, a leading sector in economic history in the West, is not given a significant role in Japan's technical modernization. On the contrary, Saigusa shows that heavy industry, beginning with the reverberatory-furnace in the 1850s and 1860s, followed by shipbuilding and railroads, had played a pioneering role in technical development in Japan.

The Second World War prevented further development of an academic history of technology modelled upon Saigusa's example. The number of articles on the history of technology in relevant journals decreased. Few other independent works using a similar pragmatic approach, with a narrower focus, were printed after Saigusa's. Okumura Shōji produced *Kōsaku kikai hattatsu-shi* (History of the development of machine tools) in 1941, and three years later *Seitetsu seikō gijutsu-shi* (Technical history of steel and iron). In 1943, Koyama Kōken (Hirotake) published a technical-historical contribution on *Nihon gunji kōgyō hattatsu-shi* (The development of the Japanese armaments industry)[14] as part of a work on *Nihon sangyō kōzō kenkyū* (Studies on Japanese industrial structure).

The breakdown in the discussion on technology shortly after 1940 indicates that political repression, based on suspected leftist activities, continued to increase. Even a pragmatic approach could not prevent repression. The upswing, which had begun with the programmatic approach by Oka Kunio in 1937 and reached a climax with the publication of Saigusa's *Gijutsu-shi* in 1940, collapsed shortly afterwards. Instead, more emphasis was placed on

[14] Koyama Kōken (Hirotake), 'Nihon gunji kōgyō hattatsu-shi' (Development history of the Japanese armament industry), in Kanbayashi Teijirō (ed.), *Nihon sangyō kōzō kenkyū* (Studies on the industrial structure of Japan), Tōkyō: Itō shoten 1943, pp. 1–148; reprinted after the Second World War (extended) Koyama Kōken (Hirotake), *Nihon gunji kōgyō no shi-teki bunseki* (A historical analysis of the Japanese armament industry), 1972, Tōkyō: Ochanomizu shobō.

the history of science, which in its self-perception also included the history of technology. According to its statutes, the History of Science Society of Japan (Nihon kagaku-shi gakkai), founded in 1941, represented the study of the development of science *and* technology (*kagaku oyobi gijutsu*...).[15] Over the following years both fields were regarded as an important part of the mobilization for war, and efforts were made to publish a multi-volume history of science (*Nihon kagaku-shi*). But this project failed to be realised, not least because of the difficulties in finding suitable authors. The series was completed only after the war.[16]

More successful was another endeavour by Saigusa Hiroto, who, following his *Gijutsu-shi*, and influenced by efforts to reappraise the history of science, undertook the task of publishing Japanese classical writings on the history of science and technology. In 1942, he began compiling the series *Nihon kagaku koten zensho* (Collected classical works on Japanese science),[17] of which ten volumes were published by 1949 (see appendix on sources of Japanese history of technology in this book).

3. POST-WAR: NEW APPROACHES IN JAPAN AND ABROAD

The post-war period saw a return of freedom of thought. Among the many intellectual currents that were now being resumed and discussed, Marxism, in many varieties, came to the fore as the strongest. It maintained this strong position until the 1970s. The discussion about the concept of technology was also reopened, following on from the pre-war debate. The theoretical approach to technology as an organization of the means of work had ultimately proved to have little potential for expansion in the history of technology and was therefore subjected to a new evaluation immediately after the end of the war. This led to new or expanded interpretations of the term 'technology'. Oka Kunio, for example, arrived at new, independent interpretations of technology, which

15 Kikuchi, 'Nihon kagaku-shi gakkai no sōritsu' (The foundation of the Japanese Society for the History of Science), p.168.

16 The result was a large monographic series edited by Nihon gakushi-in, titled *Meiji-zen Nihon kagaku-shi* (History of science before the Meiji Restoration), comprising a large number of studies on a broad range of subjects, mainly representing the history of science before the Meiji Restoration, but also including some volumes on the history of technology (see appendix on sources in this book).

17 Saigusa Hiroto (ed.), *Nihon kagaku koten zensho* (Collected works on the history of science in Japan), 10 vols., 1942–49, Tōkyō: Asahi Shinbun-sha.

he attempted to incorporate into the framework of interpersonal relationships. Starting out from the Marxist approach, he regarded production as fundamental to human society. Consequently, he wanted technology to be understood less as a cultural concept and more as production technique.[18]

Another interpretation by Taketani Mitsuo (1911–2000), physicist and Marxist philosopher of science, and Hoshino Yoshirō (1922–2007), critic of and publisher on technology (*gijutsu hyō ron-ka*), regarded technology as a 'conscious implementation of the laws of nature' (*ishiki-teki tekiyō-setsu*). In this form, the definition is intended to contribute to the solution of questions of contemporary technology as well as to the investigation of problems in historical research on technology. However, it soon became clear that such definitions did not allow for a sufficient analysis of the historical processes. Thus, in addition to these new definitions, the definition of technology as an organization of working tools remained.

The discussion about technology in the post-war period once again thrived in philosophical circles. Critics remarked that technical details were only marginally included in studies based on such concepts. In such works – according to the critics – a rather holistic approach was cultivated. One missed the 'tangible' technology in such works, which were all strongly influenced by certain ideologies. On the other hand, critics also lamented that if due space was given to technology, the socio-economic conditions, in which technology was undoubtedly embedded, were often only marginally noted.

These 'shortcomings' were the cause for the emergence of a third current of the Japanese history of technology in the post-war period. This current comprised predominantly descriptive representations of the history of technology based on a 'positivist view of history' (*jisshō-shi-kan*). This current had its origin in Kyōto in the group around the historian of technology Yoshida Mitsukuni (1921–1991). In Kyōto a positivist school of economic history had already flourished in the pre-war period and survived the Second World War almost unscathed. The approach of Yoshida Mitsukuni and his group directly or indirectly exerted a great influence on the history of technology – one that went far beyond the Kyōto group.

---

18 See Saigusa Hiroto, *Gijutsu no tetsugaku* (Philosophy of technology), Tōkyō: Iwanami 1955 (1951), pp. 268–269.

These currents in the history of technology continued to exist side by side during the post-war era. On one hand was a history of technology strongly influenced by the philosophy of technology. On the other was a more pragmatic, or even positivistic, current, dominated by industrial, economic and social history as well as the history of science and business. This dichotomy has been visible since the mid-1930s and continues to dominate academic research to this day.

4. JAPAN, MODERNIZATION AND TECHNOLOGY

While scholars in Japan resumed and further developed their pre-war discussions, a new current emerged in Western research on science and technology in Japan, which was mainly oriented towards socio-economic developments. With the upswing in the world economy after the Second World War and the decolonization that began shortly afterwards, Japan's rapid rise after its defeat during a period of high growth in the 1950s and 1960s attracted the attention of scholars around the world. The question of how Japan succeeded, and whether it could serve as a role model for developing (mostly decolonised) countries, was taken up by economists, historians, social scientists and others. The 'modernization' paradigm became a powerful strand in research, and Japan its seemingly appropriate subject. Until then, Japan's success had often been attributed to its ability to imitate developments in the West, but this did not fully explain the post-war developments. In search of the driving forces behind these changes, scholars not only looked at the rapid economic growth after the Second World War, but also at the time when Japan opened up to the West, when it was still a 'developing country' (to use a modern term that was not applied during Japan's early industrialization). In the course of the so-called modernization debate, scholars from various fields, mainly economists and social and political scientists, but also historians, both in the West (mostly the USA) and Japan, joined forces in a comprehensive research project on Japan's modernization and its legacy from the Edo period. Their efforts resulted in a series of six books, published by Princeton University Press, which formed an influential counterpoint to the Marxist perspective prevailing in Japan.[19] The 'modernization

[19] The Princeton series under the title 'Studies in the modernization of Japan' (Princeton University Press) consisted of six volumes, the authors of which

project' and the Princeton series received considerable attention in Western countries and in Japan and, together with many similar publications, left its mark on Japanese studies for quite some time. Knowledge on Japan's industrialization, and its socio-economic and political development, which had been scarce abroad, was rapidly accumulated and widely published.

Astonishingly enough, despite the large bulk of studies, research on technology or technological development remained scarce. In numerous publications on Japan's economic and industrial development, the terms 'technology' or 'technique' either do not appear at all, or only in passing. While industry, capital formation, institutional structures, bureaucracy, education, etc., were treated in detail, technology was rarely considered worth studying.

An exception was the question of technology transfer from the West to Japan, which became a major topic in the 1970s, and was also projected back into history. Explanations identified the import of foreign machines and equipment as well as technical experts (engineers, technicians, technical workers, etc.) as important agents that brought Western technology to Japan. Another group of transmitters were Japanese who had been sent abroad to study. They are said to have conveyed the technical knowledge that Japan needed for its rise.

These narratives depict technology transfer mostly as a one-way street, and rarely or not at all address the question of the conditions necessary for a successful implementation. It is often overlooked that a technical development – and industrialization is a 'technical' development – cannot start from scratch and that successful technology transfer always requires a certain base in the recipient country. There must be some technical education, knowledge and understanding to enable the transfer and implementation of foreign technology. However, neither the technical knowledge and production processes developed in Japan itself, nor the complex and multi-level technical and scientific education system established with the Meiji Restoration, were addressed in the technology transfer studies.

---

regarded Japan's modernization mostly as a process converging towards the US/Western model of a democratic society and market economy. The volumes were: Marius B. Jansen (ed.), *Changing Japanese Attitudes toward Modernization* (1965); William E. Lockwod (ed.), *The State and Economic Enterprise in Japan* (1965); Robert E. Ward (ed.), *Political Development in Modern Japan* (1968); Ronald P. Dore (ed.), *Aspects of Social Change in Japan* (1971); Donald Shively (ed.), *Tradition and Modernization in Japanese Culture* (1971); James Morley (ed.), *Dilemmas of Growth in Prewar Japan* (1972).

This only changed in the 1980s and 1990s. It was certainly the influence, or rather the aftermath, of the various development theories that emerged during the modernization debate, that in the late 1970s led to a project initiated by the United Nations University in Tōkyō. This was carried out by the Institute of Developing Economics under the direction of Hayashi Takeshi, and was entitled 'Technology Transfer, Transformation, and Development: The Japanese Experience'. The aim and target audience of this project are stated in the introduction:

> The United Nations University, in commissioning this project on the Japanese experience, observed that Japan, once an importer of modern technologies from the West, has now developed itself to the point of being an exporter of the latest technologies, and, further, that what it was that made such a transformation possible was a matter of great interest to developing countries.[20]

Hayashi continues by embracing the voices of the developing world:

> Japan has become the focus of attention among the developing countries, and I find this interest perhaps may have great practical value.

On the basis of such considerations, numerous renowned Japanese historians prepared a wealth of working papers (80 papers and two booklets) which with an emphasis on technology transfer and technical development, traced the course of Japanese modernization in many individual areas such as iron and steel, mining, textiles, etc. Many of these papers were presented at conferences and ultimately published as books.

As valuable as the project was, and as interesting and detailed as the individual studies were, they only partially contributed to the history of technology in Japan in an overarching sense because the history of technology was pushed into the background. Economic history was at the forefront. In retrospect, this project seems to have marked the end of modernisation studies and a whole period of development studies. It also had no lasting influence on the broad history of technology in Japan. It is only today that this

[20] Hayashi Takeshi, *The Japanese Experience in Technology: From Transfer to Self-reliance*, Tōkyō: United Nations University Press, 1990, Part I, Overview: Chapter 2 passim.

project is being remembered again and that the publications are again being used.

Another attempt to introduce the field of the Japanese history of technology outside Japan was carried out at the Japan Center of the University of Marburg, Germany, in the 1990s. It brought together Japanese historians of technology, who contributed articles in English on their respective research topics to provide a first, albeit still not comprehensive, overview of the various topics in the Japanese history of technology.[21]

In the two decades before the turn of the millennium, Western scholars such as James Bartholomew[22], Morris Low[23] and especially Tessa Morris-Suzuki[24] also published major studies on this subject, which became important milestones in this strand of intensifying research on Japan's technological development and its historic roots.

Despite such publications, however, Japanese Studies did not succeed in taking the position of the history of technology out of its niche existence and establishing it as a subject in its own right, as has been the case with the history of science and technology focusing on China. Japanology in Europe has traditionally been rooted in philological studies in fields like literature, religion and history, and while social science-related topics have gradually gained ground in Japanese studies since the 1970s, other subjects such as the history of technology have attracted less attention. Publications in these fields by scholars dealing mainly with Japan on the basis of Japanese sources remained largely the exception.[25]

---

21 The resulting three-volume publication finally comprised 31 contributions that ranged from Japan's early history to the present. See Erich Pauer (ed.), *Papers on the History of Industry and Technology in Japan,* (vol. I: From the Ritsuryō-system to the early Meiji-Period; vol. II: From the Meiji-Period to Postwar Japan; vol. III (ed. with Sakata Hironobu): History of Glass in Japan), Marburger Japan-Reihe, vol. 14/1 – 14/3, Marburg 1995.

22 James R. Bartholomew, *The Formation of Science in Japan: Building a Research Tradition*, Yale UP 1993.

23 E.g., Morris Low, Nakayama Shigeru, Yoshioka Hitoshi (eds), *Technology and Society in Contemporary Japan,* Cambridge: Cambridge UP 1999.

24 Tessa Morris-Suzuki, *The Technological Transformation of Japan: From the Seventeenth to the Twenty-first Century*, Cambridge: Cambridge UP 1994.

25 New interdisciplinary orientations such as STS (Science and Technology Studies or Science, Technology and Society), which have flourished in recent decades, gained a foothold in Japan at the turn of the millennium and are now widespread, but they are a field of their own, that cannot be considered in detail here.

5. IN THIS BOOK

A new attempt at a more intensive study of the Japanese history of technology in the West was made in the early years of the new millennium. This was a research project with the title 'From craftsman to engineer', planned by the editor of this book to focus on the people or groups involved in the development of technology from the Edo period to the Meiji period and beyond. The aim was to highlight the transition period in particular from a time in which technical knowledge was – e.g., in the crafts – primarily passed on orally, through the appearance of the first written sources, to the beginnings of technology transfer from the West and its reception by a new but rather small stratum of technically trained personnel, and finally to a broadened, formalised technical education along Western lines, which brought forth the 'engineer' in the Western sense in Japan.

Owing to various obstacles, the realisation of this project was delayed. It was only in 2018, that the opportunity arose to make the first step and initiate a conference entitled 'Generation and dissemination of technical knowledge in Japan from the Edo- to the Meiji-Period'. The aim was not to look back at the technical developments of the Edo period from the perspective of industrialization, but to examine how in a period of more than 200 years an independent cluster of various technologies emerged. The main focus was on the processes of how mostly indigenous, as well as imported technical knowledge (in a broad sense), was generated, disseminated and applied in the Edo period. It was a further goal to track down traces of the process of the transition from an orally transmitted (often secret) professional knowledge to a more publicised means of distribution in networks or even in printed form, reflected in the emerging infrastructure providing for basic education and a print and book market. Some of the presentations at this conference became the basis for the chapters in this book.

Going back in Japanese history, one question was how knowledge, indigenous or imported, was transmitted. The absence of written records on technical issues in early Japanese history allows only the conclusion that this knowledge was passed on nearly exclusively orally. A great change becomes visible in the late fifteenth and early sixteenth centuries. The decentralization of power and the emergence of strong domains, the intensification of trade with China in the fifteenth century, and the landing of the Portuguese in Japan in 1543, acted as stimuli in many areas and also led to the introduction of new technical knowledge, methods and devices. In the course of the wars waged by Toyotomi Hideyoshi in Korea at the end of the sixteenth century, Korean craftsmen were forcibly brought to Japan, where they orally spread their superior knowledge in many crafts and trades, above all porcelain production, among Japanese craftsmen. Shortly afterwards, a growing number of written documents came to Japan, imported from China, not least agricultural textbooks, which were important for Japan and her wish for self-sufficient food supplies. They became the foundations of a new knowledge based on technology transfer via written documents, i.e., books.

The import of Chinese writings in many different fields contributed to the establishment of a book and print culture in Japan. Besides letterpress printing, which flourished briefly in early seventeenth century, woodblock printing opened up new possibilities. The rapid increase of woodcut-printed books and the trade in books affected not only the field of fine literature, but also the dissemination of knowledge, including technical knowledge. This opened a new chapter in the generation and dissemination of knowledge in general and also in the field of techniques. But in the course of the seventeenth century something special was added: illustration.

Illustrations in literary works, like books and picture scrolls, were not new for the Japanese in the seventeenth century, but the way in which they were increasingly incorporated into works of non-fiction, into publications obviously intended as 'textbooks', opened up a new form of passing on 'visualised' knowledge. This was facilitated by woodcut printing which allowed the combination of text and pictures without difficulty.

With this, Japan was entering an epoch in which knowledge was increasingly conveyed in the form of pictures. Providing information that had previously been only conveyed as text also through illustrations was a development in book production that spread rapidly. Initially texts were supplemented with a few small pictures, but the written word remained the dominant feature. In some subject areas, however, especially for self-study, the relationship between text and pictures was soon reversed in the course of the Edo period. Pictures were now combined with the text, and soon entire pages were designed and printed to make illustrated books. Another genre was picture scrolls on gold and silver mining, which often had little or hardly any text (see Mathias in this volume), but gave detailed depictions of the mining and smelting processes.

An interesting and early example of the transmission of knowledge through images is an illustrated encyclopaedia entitled *Kashiragaki zōho kinmō zui* by Nakamura Tekisai (1629–1701), published in multiple editions from the seventeenth century onward. It contains not only large representations and explanations of animals, plants, etc. but also, for example in volume 10, equipment for agriculture, the processing of agricultural products (such as mills and presses), and the processing of cotton and the production of threads, etc. (such as looms). It was an encyclopaedia, not a textbook, and the illustrations were not yet intended to reproduce the equipment shown for practical use. But this book was a kind of forerunner of later books with pictorial instructions for making the devices.

Using illustrations to bring complex information closer to the reader is of great advantage, especially in teaching. Facts that are difficult to convey as text can be simplified and, especially against the background of already existing knowledge, can often be disseminated more precisely in pictorial form.

This is clearly shown in the contribution by Annick Horiuchi, when in the *Nippon sankai meisan zue* (Famous products of Japan from mountain and sea, illustrated) the pictures show not only certain objects (tools, etc.), but also the different steps in particular production techniques in a single illustration. The same is shown in Pauer's article in this volume on early agricultural treatises, which later were expanded to a kind of illustrated catalogue of various tools of a certain type. Another kind of illustration is taken up in the article by Itō Mamiko, which shows a crowd at an exhibition, with a group sitting before a human skeleton, a person who is obviously giving explanations, and a written text in the background.

The use of illustrations, not detailed but in a very sketchy form, was particularly important for craftsmen such as carpenters and joiners. In these trades it was common to pass on working methods only within the family, as well as to apprentices coming from outside. In making more complicated tools, such as those used in the Edo period, such as in silk spinning, agriculture, etc., the written word was not enough, if only because not all apprentices could be expected to be able to read and write. Pictorial representations were required here. They could be copied and passed on, using workshop books or sample books with keywords or short explanations of the – mostly rough – technical sketches, as shown in the contribution by Pauer.

Such pictures or illustrations could be read intuitively and then put into reality. This development must be seen as particularly important for the dissemination of knowledge in the Edo period. The first contributions in this volume (Horiuchi, Pauer, Mathias) provide a number of examples in this respect. It was not until the end of the eighteenth century that this way of passing on knowledge was changed. At that time the first textbook on the manufacture of watches came onto the market, thus opening the era of technical textbooks in the current sense.

Whereas the first four contributions in this volume deal with developments in the Edo period, the last three essays reflect the beginning of the transition from the Edo to the Meiji period. The paper by Hashimoto Takehiko deals with this transition period and shows the technical level that craftsmanship (especially in metalworking) had reached in the Edo period.

Tanaka Hisashige, known also as the 'Karakuri Giemon', was an ingenious craftsman who manufactured not only automata (mechanic puppets, *karakuri ningyō*) but also clocks and other devices. He later became the founder of a metalworking business that was the forerunner of today's Toshiba.[26] The contribution by Hashimoto Takehiko, who observed the process of dismantling of Tanaka Hisashige's so-called 'myriad year clock' and the building a replica, presents a comprehensive, detailed description of the clock. Looking at its gears, axles and springs, one is stunned by the accuracy with which these parts have been manufactured. The clock is also a prime example of precise metalworking in brass, which reached a peak at this time. One must not forget, however, that the clock also makes visible detailed knowledge and understanding of the celestial bodies, their movement and changes in the course of the seasons. The clock also embodies arithmetical calculations: the gear ratios had to be computed before they were made. Manufactured in the middle of the nineteenth century, the clock is, even today, an engineering marvel.

A Cohort of craftsmen in the years shortly after the Meiji Restoration (1868) is analysed by Nishiyama Takahiro, based on information from a survey of a particularly outstanding group as an example. The survey, *Tōkyō meikō kagami* (Highly skilled craftsmen of Tōkyō), was conducted in 1875. It gives insights into aspects of

---

[26] For Tanaka Hisashige see Imazu Kenji, *Karakuri Giemon – Tōshiba no sōritsusha Tanaka Hisashige to sono jidai* (Karakuri Giemon – The founder of Tōshiba and his time), Tōkyō: Daiyamondo-sha 1992.

traditional craftsmanship in Tōkyō, craftsmen's main places of residence, apprenticeship period, establishment of businesses, the number of employees, etc. It hints at the transitions in these crafts under the changed circumstances of the early Meiji period.

This analysis provides the background of Suzuki Jun's contribution on the boilermaking industry in the late nineteenth century. The factories, where boilers for industries, such as silk-reeling businesses, machine factories, and mines, were built, relied heavily on the craftsmen who are the subject of Nishiyama's article.

Even though steam and its production were well known in the Edo period, there was no development towards the use of steam power in production processes. For this, the transfer of technology from abroad was necessary. Suzuki clearly shows the different paths taken in this process. An important part of his research is the analysis of the establishment of a system of boiler inspections in the Meiji period, by which boiler usage was made as safe as possible.

Altogether, the conference and the essays collected in this volume highlight important technological developments in early modern Japan. In addition to the still widespread oral transmission of empirical technical knowledge, texts and images increasingly appeared as transmitters of knowledge in crafts and trades. This promoted not only the range and speed of dissemination of knowledge, but also the accumulation of knowledge in specific sectors such as agriculture and metalworking.

The few examples in this book shed a little light on the technical development and dissemination of knowledge in the Edo period. Numerous other chapters of the Japanese early modern history of technology still await in-depth study.

It is hoped that this work will stimulate historians of technology and representatives of other disciplines to broaden their view and include Japan and other countries beyond Europe and North America as a matter of course in their reflections on technological development.

REFERENCES

Aikawa Haruki, *Gendai gijutsu-ron* (On contemporary technology), Tōkyō: Mikasa shobō,1940.

Bartholomew, James R., *The Formation of Science in Japan: Building a Research Tradition*, New Haven, Conn., Yale University Press, 1993.

Chūjō Yuriko, "Rajio ogon jidai' no teichō' (The weakness of the 'Radio's golden age'), in *Yuibutsu-ron kenkyū*, no. 58, 1937, pp. 34–40.

Hayashi Takeshi, *The Japanese Experience in Technology: From Transfer to Self-reliance*, Tōkyō: United Nations University Press, 1990.

Imazu Kenji, *Karakuri Giemon – Tōshiba no sōritsusha Tanaka Hisashige to sono jidai* (Karakuri Giemon – The founder of Tōshiba and his time), Tōkyō: Daiyamondo-sha, 1992.

Kikuchi Toshiyoshi, 'Nihon kagaku-shi gakkai no sōritsu' (The foundation of the Japanese Society for the History of Science), in *Kagaku-shi kenkyū* (Journal of history of science, Japan), vol. 32 (187), 1993, pp. 166–173.

Koyama Kōken (Hirotake), 'Nihon gunji kōgyō hattatsu-shi' (Development history of the Japanese armament industry), in Kanbayashi Teijirō (ed.), *Nihon sangyō kōzō kenkyū*, Tōkyō: Itō shoten, 1943, pp. 1–148; reprinted after the Second World War (extended) Koyama Kōken (Hirotake), *Nihon gunji kōgyō no shi-teki bunseki* (A historical analysis of the Japanese armament industry), Tōkyō: Ochanomizu shobō, 1972.

Low, Morris, Nakayama Shigeru, Yoshioka Hitoshi (eds), *Technology and Society in Contemporary Japan,* 5 vols. + 1, Cambridge: Cambridge UP 1999.

Morris-Suzuki, Tessa, *The technological transformation of Japan: From the seventeenth to the twenty-first century*, Cambridge: Cambridge UP, 1994.

Oka Kunio, 'Gijutsu no rekishi-teki kenkyū – Gijutsu-shi-ron josetsu' (The historical study of technology – Introduction to the theory of the history of technology), in *Chūō kōron* no. 7 (no. 596), 1937, pp. 52–61.

Oka Kunio, 'Gijutsu-shi ni tsuite – Gijutsu-sho joron ho-i – jō/ge' (On the history of technology – An amendment to the introduction to the history of technology – parts one and two), in *Yuibutsu-ron kenkyū* no. 57, 1937, pp. 19–26 and no. 58, 1937, pp. 142–148.

Oka Kunio, 'Waga kuni ni okeru denki gijutsu no hattatsu' (The development of electrical engineering in Japan), in *Yuibutsu-ron kenkyū*, no. 52, 1937, pp. 110–132 (Part I), no. 53, 1937, pp. 132–149 (Part II), no. 55, 1937, pp. 106–122 (Part III), no. 56, 1937, pp. 114–134 (Part IV).

Okumura Shōji (pseud. Ishikawa Yoshikazu), 'Dōgu kara kōsaku kikai e no hattatsu' (The development from tools to machine tools), in *Yuibutsu-ron kenkyū*, no. 38, 1935, pp. 106–114.

Pauer, Erich (ed.), *Papers on the history of industry and technology in Japan,* (vol. I: *From the Ritsuryō-system to the Early Meiji-Period*; vol. II: *From the Meiji-Period to Postwar Japan*; vol. III (ed. with Sakata Hironobu): *History of Glass in Japan*), Marburger Japan-Reihe, vol. 14/1–14/3, Marburg 1995.

Saigusa Hiroto (ed.), *Nihon kagaku koten zensho* (Collected works on the history of science in Japan), 10 vols., Tōkyō: Asahi Shinbun-sha, 1942–49.

Saigusa Hiroto, *Gijutsu-shi* (History of technology) (Gendai Nihon bunmei-shi, vol. 14), Tōkyō: Tōyō keizai shinpō-sha, 1940. Reprinted in *Saigusa Hiroto chōsaku-shū* (Collected works of Saigusa Hiroto), vol. 10, pp. 13–288, Tōkyō: Chūō kōron-sha, 1973.

Saigusa Hiroto (ed.), *Nihon kagaku koten zensho* (Collected works on the history of science in Japan), 10 vols., Tōkyō: Asahi Shinbun-sha, 1942–49.

Saigusa Hiroto, *Gijutsu no tetsugaku* (Philosophy of technology), Tōkyō: Iwanami, 1955 (1951).

Satō Kazuo, 'Nihon ni okeru kōsaku kikai no saikin no hattatsu' (On the recent development of machine tools in Japan), in *Yuibutsu-ron kenkyū*, no. 57, 1937, pp. 77–93.

Tawara Kuniichi, *Korai no satetsu seiren-hō (Tatara-buki seitetsu-hō)* (Traditional iron-smelting (The *tatara-buki* smelting technique)), Tōkyō: Maruzen, 1933.

Tsuruta Michio, 'Meiji-jidai ni okeru waga kuni kōsaku kikai kōgyō hatten no tokushitsu' (The peculiarity of the development of the machine tool industry in Japan in the Meiji era), in *Gakugei*, no. 67, 1938, pp.124–33.

Yamazaki Toshirō, 'Japan's Contribution to the Modern History of Technology', in *Japanese Studies in the History of Science* 1, 1962, pp. 45–47.

1

Production Techniques in Early Modern Japan as seen through 'Famous Products of Japan from Mountain and Sea, Illustrated' (*Nippon sankai meisan zue*, 1799)

Annick HORIUCHI

1. INTRODUCTION

FOR ECONOMIC HISTORIANS of early modern Japan, the second half of the Tokugawa era is marked by a rapid growth in commercial production in rural areas. In the seventeenth century, it was already not uncommon for farmers to engage in production activities in addition to the (mandatory) cultivation of rice. However, their products were mainly intended for domestic consumption and were not marketed. In the central region around Ōsaka and Kyōto, commercial activities took off in the rural areas which supplied these cities with the raw materials they needed for their quality craftsmanship. To meet ever-increasing demand, production techniques were improved and rationalized. It became the norm to specialize in a few specific products and, from the mid-eighteenth century, this became the case across the entire country. Most provinces were involved in some sort of specialized production involving significant numbers of workers. Fresh foods such as fish or vegetables, processed foods such as oil, soy sauce, *sake*, and dried fish, or raw materials for textile production such as cotton or raw silk were produced in great quantities in rural areas and marketed locally or nationwide.[1] By the end of the century, there was no longer any province that did not have its own 'famous products' (*meisan* or *meibutsu*).

Nippon sankai meisan zue ('Famous Products of Japan from Mountain and Sea, Illustrated'), hereafter *Meisan zue*, first printed in 1799 in Ōsaka, sheds light on the scale and complexity of production techniques that had been achieved by the end of the eighteenth century. Through its detailed illustrations, *Meisan zue* shows the new face of the Japanese countryside. The population of Ōsaka, who were familiar with these 'famous products', could for the first time have an insight into the production environment. The book was evidently a great success. It was reprinted several times before the Meiji era, as shown by the large number of copies held today in Japanese, as well as in European, libraries.[2] More surprisingly, many pictures from the book appear, with only minor changes, in the *Dai Nippon bussan zue*,[3] a catalogue of a hundred multi-coloured prints representing scenes of production for the first Exhibition for the Promotion of National Industries (*Naikoku kangyō hakurankai*) held in 1877.

This paper aims to resituate *Meisan zue* in the intellectual context of the late eighteenth century. I argue that the book was intended to be read by a scholarly audience and can be placed in a tradition of studies that had been developing since the early seventeenth century. Focusing on the way production techniques are described and represented, I attempt to determine how production activity was seen by scholars, the message they wanted to convey through their detailed descriptions and, more generally, the impact this endeavour might have had in light of the period and the place it was published. Before focusing on these questions, we shall start by identifying the salient features of *Meisan zue*.

---

1 See Shinpo Hiroshi and Hasegawa Akira, 'Shōhin seisan ryūtsū no dainamikusu' (Dynamics of the production and circulation of commercial goods), in Hayami Akira and Miyamoto Mataji (eds), *Keizai shakai no seiritsu 17–18 seiki* (Development of economy and society in the seventeenth to the eighteenth century) (Nihon keizai 1), Tōkyō: Iwanami shoten, 1988, pp. 217–270.

2 This may be seen, for example, in the Union Catalogue of Early Japanese Books Database of the National Institute of Japanese Literature (http://base1.nijl.ac.jp/infolib/meta_pub/G0001401KTG). As of April 2019, we know that at least eighty-two copies are preserved in Japanese and Western libraries, in addition to the copies listed in the *Kokusho sōmokuroku* (General catalogue of national books). This list is far from complete.

3 The book was printed in Tōkyō in 1877 (Meiji 10), with illustrations by Utagawa Hiroshige III.

2. THE BOOK

Nippon sankai meisan zue is a medium-sized book composed of five fascicles of twenty to forty folios each. No author's name is mentioned on the title page and the colophon indicates only the name of the illustrator, Shitomi Kangetsu, a successful painter in Ōsaka during the second half of the eighteenth century. Here we shall leave aside the issue of the author's identity and focus only on the book and its characteristics.[4] The preface was written by a well-known Ōsaka-based scholar by the name of Kimura Kenkadō (1736–1802), who had a reputation as an expert in natural history (Jap. *honzō*; Ch. *bencao*),[5] painting, calligraphy, ancient and contemporary books from Japan, China and Europe, as well as other fields.[6]

Meisan zue was written for a highly educated audience, as can be seen by the frequent references to Chinese and Japanese classics, and the writing style which uses many Chinese terms. However, by the end of the eighteenth century the readership of this book genre had grown considerably. Publishers took this widened readership into account, systematically adding *furigana*, i.e. phonetic glossings

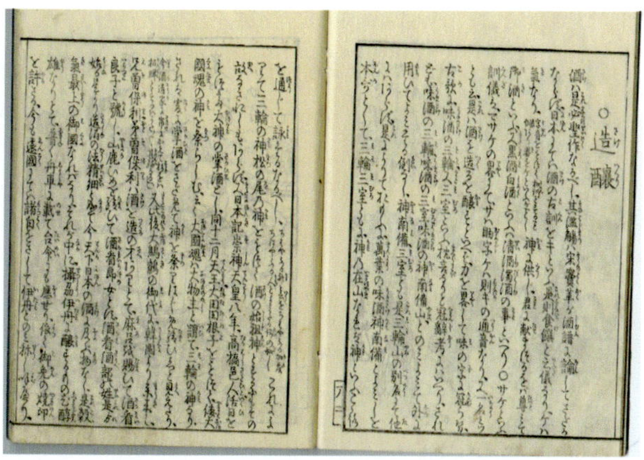
○造釀

Fig. 1: Section 'The making of *Sake*', *Nippon sankai meisan zue*, fasc. 1. Source: Castillon collection. Collège de France.

4 See Higuchi Hideo's introduction (*kaisetsu*) in the facsimile edition of the book. *Nippon sankai meisan zue*, Tōkyō: Meicho kankōkai, 1979, pp. 1–11.

5 See below for an explanation of what is meant by *honzō* studies or *honzōgaku*.

6 Ōsaka Rekishi Hakubutsukan ed., *Kimura Kenkadō: naniwa chi no kyojin* (Kimura Kenkadō: A giant of Ōsaka's learning), Kyōto: Shibunkaku shuppan, 2003.

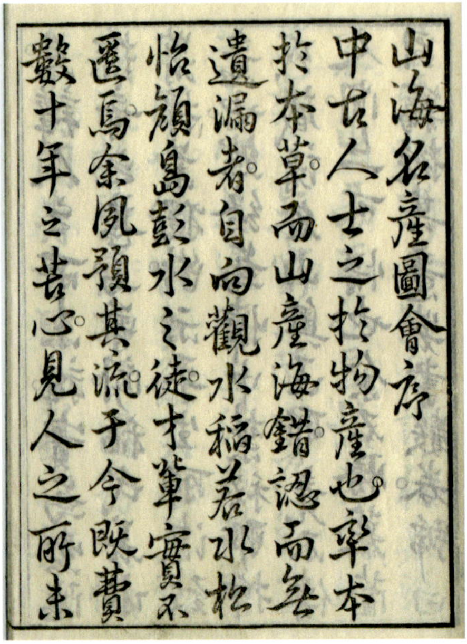

山海名産圖會序
中古人士之於物産也率本
於本草而山産海錯認而無
遺漏者自向觀水稲若水松
怡顔齋彭水之徒才輩實不
遑焉余夙預其流于今既費
數十年之苦心見人之所未

Fig. 2: *Nippon sankai meisan zue*, Preface.
Source: Castillon collection. Collège de France

of *kanji* (see Fig. 1). As Kimura Kenkadō noted in his preface, this made the book accessible even to "women and children."

Kimura Kenkadō's preface, written entirely in classical Chinese (see Fig. 2), sheds light on the tradition of learning in which he placed the book. He established a close relationship between its content and *honzōgaku* (study of *materia medica*) tradition into which he himself had been initiated. *Honzō* (Ch. *bencao*) or *honzō gaku* had undergone spectacular development in Tokugawa Japan. Though already well-established as a field of knowledge in medieval times, the status of *honzōgaku* rose considerably with the importation to Japan of *Bencao gangmu* (*Compendium of Materia medica*) at the beginning of the seventeenth century.[7]

Printed in 1596, *Bencao gangmu* is an encyclopaedic work of fifty-two fascicles, containing more than 1,800 entries, arranged under sixteen sections (*bu*) and sixty categories (*lei*). It provides a wealth

---

7 Yabe Ichirō, *Edo no honzō: yakubutsugaku to hakubutsugaku* (The *materia medica* of Edo: from pharmacopoeia to natural history), Tōkyō: Saiensu-sha, 1984, pp. 25–42.

of information on plants, minerals and animals, and was synthesized from a very large number of ancient and contemporary books, as well as from personal observations. Special attention is paid to identifying as closely as possible the names of substances, places of production, mentions of the product in literary works, morphology, flavours, animal behaviours, and so on. This clearly shows the author's ambition to practice 'the investigation of things' as advocated by Confucian scholars.[8] At the same time, the book retains a strong medical orientation through the many medicinal recipes listed in each entry.[9]

Bencao gangmu went through many editions in the archipelago. Very quickly, Japanese scholars' directed their attention to substances (herbs, trees, minerals, fish, birds, etc.) that could be found or grown in Japan. At the turn of the eighteenth century, two works, *Yamato honzō* (*Materia medica of Japan*)[10] and the *Honchō shokkan* (*Mirror of food in our empire*)[11] showed that Japanese scholars had begun to build their own tradition, rooted in the lands of Japan. Neither of them contains medicinal recipes. Both of them concentrate on names, appearances in (Chinese and Japanese) literature, morphology, animal behaviors, tastes, geographical distribution, and modes of consumption, with the strictly medical content of the entries drastically reduced.

At the same time, because of shogunal policy and financial difficulties in various domains, *honzōgaku* specialists were strongly encouraged to develop domestic products (natural or manufactured, ancient or new, of Japanese or foreign origin) that could be profitably marketed. *Honzōgaku* was thus transformed into *bussangaku* or the 'science of products', a field partly driven by mercantile considerations.[12] *Meisan zue*, with its emphasis on famous, high-quality products, is emblematic of this transformation. However, as

8 Métailié Georges, 'The *Bencao gangmu* of Li Shizhen: an Innovation in Natural History?' in E. Hsu (ed.), *Innovation in Chinese Medicine*, Cambridge: Cambridge University Press, 2001, pp. 223–224.

9 Métailié notes, 'The compilation of these recipes should not be underestimated, for it makes the *Bencao gangmu* a precious reservoir of practical medical knowledge accessible to physicians in the late Ming.' Métailié Georges, 'The *Bencao gangmu*,' p. 252.

10 Kaibara Ekiken's *Yamato honzō* (1709) was the first *honzō* treatise written in Japanese dedicated to the natural species found in Japan.

11 Hitomi Hitsudai's *Honchō shokkan* (1697) focuses on natural species that can be consumed as food. The book is written in classical Chinese.

12 Historians of *honzō* generally consider that *honzō* specialists were strongly committed to the observation of nature and less interested in the the medical use of substances in comparison to their Chinese counterparts. See Yabe Ichirō, Edo no *honzō*, pp. 79-91.

will be discussed, it preserves many features of the *honzō* tradition, particularly its interest in identifying the correct names of products, discussing differences according to regions of production, or tracing past references to products.

Unlike *Bencao gangmu*, *Meisan zue* is far from exhaustive or systematic. The Japanese term *zue*, translated here as 'illustrated', only means that the book was designed as an illustrated album, where products were described with pictures.[13] The fifty entries in the book are of unequal length, and the selection criteria are not made explicit. This could be because *Meisan zue* was not, strictly speaking, a new editorial project. It was designed as a sequel to another book, *Nippon sankai meibutsu zue* (hereafter *Meibutsu zue*), published more than forty years earlier in 1754 by Hirase Tessai (?–?).[14]

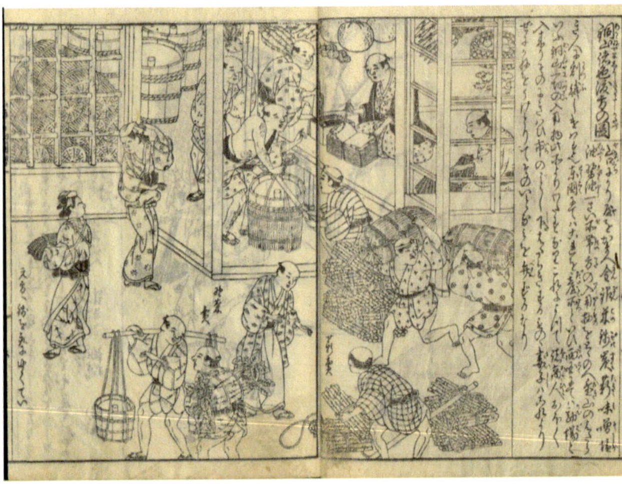

Fig. 3: *Nippon sankai meibutsu zue*, fasc. 1 (1754).
Source: National Institute of Japanese Literature, ID. 200021849 https://kotenseki.nijl.ac.jp/biblio/200021849/viewer/10

[13] The publication of *Meibutsu zue* and *Meisan zue* coincides with the period where books called *Meisho zue* enjoyed great success. *Meisho zue* means 'famous places, illustrated'. For further detail, see Shirahata Yōzaburō, 'The Printing of illustrated travelogues in eighteenth century Japan', in Susanne Formanek & Sepp Linhart (eds.), *Written texts – Visual Texts Woodblock-printed Media in Early Modern Japan*, Leiden: Hotei Publishing, 2005, pp. 199–214.

[14] According to Higuchi Hideo, *Meisan zue* was designed as a sequel to *Meibutsu zue* and written by the same author. See Higuchi's introduction in the facsimile edition of the book, *Nippon sankai meisan zue*, Tōkyō: Meicho kankōkai, 1979.

Meibutsu zue and *Meisan zue* share many commonalities. The titles, which differ only by a single Chinese character, have almost the same meaning. Both focus on a selection of regional products and provide vivid and accurate descriptions of the production context. In both cases, the illustrations are an essential complement to the text. However, there was a span of forty-five years between them, which explains why they diverge significantly on some points. First, as Chiba Tokuji notes, the selected products are of a different type. While agricultural and forest products from the Kinki region are predominant in *Meibutsu zue*, seafood products (e.g. fish, shellfish) or products from mountains or 'distant lands' (e.g. Ezo, Kyūshū, foreign countries) occupy the bulk of *Meisan zue*.[15] Second, each entry in *Meibutsu zue* was designed according a single pattern, composed of a short six-line text with a picture. In *Meisan zue*, entries are of a variable length. They usually give a quite detailed description of the production process. Its seventy-one plates, each of which cover two whole pages, are not only beautiful, they also provide insight into the natural, human and technical environment in which the production took place (see Fig. 4). Moreover, they reveal the illustrator's talent in capturing the scene of production in all its detail.

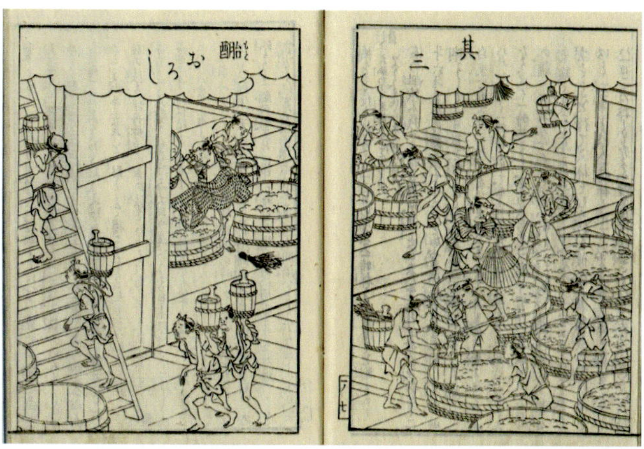

Fig. 4: '*Sake* production in Itami', *Nippon sankai meisan zue*, fasc. 1.
Source: Castillon collection. Collège de France.

[15] See Chiba Tokuji (ed.), *Nippon sankai meisan meibutsu zue*, Tōkyō: Shakai shisōsha, 1970, pp. 300–301 and the table of contents of the *Meisan zue* at the end of the present article.

While *Meisan zue* says nothing about the author's motive in producing the work, *Meibutsu zue* is more explicit. The postface, written by the author's son, reveals that the intent was to showcase production sites that were not well known to the public and that plates had been drawn from 'direct observation' onsite for this purpose. These remarks would also be true for *Meisan zue*, whose plates were like snapshots taken on the spot. On this point, the *Meisan zue* was a typical publication of its time. Many publications indeed contained faithful representations and bird's-eye views of landscapes or famous sites. The illustrator and/or the author considered it their duty to observe things themselves and to transmit what they had seen. The distance between the reader and the production site was thus considerably reduced. These points may be illustrated by a detailed consideration of two examples: *sake* brewing and tuna fishing, both of which are representative of the *Meisan zue* style.

3. TWO EXAMPLES : SAKE BREWING AND TUNA FISHING

3.1. Sake brewing

The first volume of *Meisan zue* is devoted entirely to describing *sake* brewing in Itami, located today in Hyōgo prefecture.[16] *Sake* brewing had been practiced in Japan since ancient times, but from the end of the sixteenth century and during the seventeenth century its production underwent substantial improvements in the Ōsaka region.[17] Itami was one city that gained fame in the seventeenth century by developing high-quality *sake*. Itami *sake* was particularly appreciated in the capital city of Edo, to which most of it was exported.

Sake production was a long and delicate process, the success of which depended greatly on the quality of its ingredients (rice, water, and *kōji*) and the *savoir-faire* of its master brewer (called *tōji*).[18] Each house generally kept its production techniques secret.

---

16 For the topics covered in the *Meisan zue*, see the table at the end of the article.

17 *Meisan zue* notes, 'The number of wine brewers has continued to grow, and nowadays, the wine brewed in Itami, Ikeda, and elsewhere in the same province in Nishinomiya, Hyōgo, Nada, and Imazu is an excellent product.' (translation by W. Boot).

18 Yunoki Manabu, *Sakezukuri no rekishi* (History of *sake* making), Tōkyō: Yūsankaku, 2018 (first printed in 1975 as *Nihonshu no rekishi*).

It was therefore unusual to reveal, as *Meisan zue* does, all the manufacturing secrets. It is said in the book that the method described was 'the method of *only one* house in the village of Itami. The others have each their own secrets.' One can imagine that the author or the illustrator was an acquaintance of the brewer, who accepted to disclose some of his know-how.[19]

In the description of the procedure given in the *Meisan zue,* four main steps can be identified, corresponding to the making of four mixtures or ingredients:[20]

1) *moyashi*: a mold (*aspergillus oryzae*) obtained from old rice and ashes of zelkova wood, used to convert rice starch to fermentable sugars.[21]
2) *kōji*, obtained from a quantity of steamed rice mixed with a small amount of *moyashi* in a ratio of approximately two *go*/ (360 ml) of *kōji* to one *koku* (180 *l*) of steamed rice.
3) *moto* (the 'base'), a thick paste obtained from a first fermentation of fresh steamed rice mixed with water and *kōji* in fixed proportions.
4) *moromi*, the *sake* in a non-clear liquid state, obtained from a second fermentation process, resulting from three successive additions (*soe*) of steamed rice, water and *kōji* to the previous mixture, each addition being carried out under fixed conditions.

The clear *sake* obtained after filtration and pasteurization is put in barrels to be marketed.

Here is an excerpt that describes the making of *kōji*:

> *Sake kōji* (Because it was formerly made from wheat, it was written using the character 麹 with the wheat radical. The Chinese process is extremely complicated, but the Japanese method is easy).

19 It is tempting to relate this proximity to the fact that Kimura Kenkadō had a license as a *sake* brewer (*kabu*). See Ōsaka Rekishi Hakubutsukan (ed.), *Kimura Kenkadō: naniwa chi no kyoji*, Footnote 6, p. 22.

20 In the book, the entry is divided into the following parts: 1) *sake kōji* 2) *moyashi* (rice mold) 3) *moto* 4) Additions (*soe*) 5) Rice [selection] 6) Washing of rice 7) Particular traditions 8) Necessary utensils 9) Rectification ashes 10) Rice alcohol (*mirin*) 11) Rice vinegar 12) Washing of rice 13) Atago Festival.

21 "Sake". Britannica Academic, *Encyclopædia Britannica*, 4 Jan. 2019. academic-eb-com.rproxy.sc.univ-paris-diderot.fr/levels/collegiate/article/sake/65000. Accessed 22 Mar. 2019.

Around the equinox, four days before the appointed day when the 'base' (*moto*) will be put in, they wash the rice in the morning and let it soak in water for one day. The next day, they steam the rice and spread the cooked rice on bamboo mats. Using a rake, they mix the rice and level it. At the moment when the rice reaches skin temperature, they throw all of it into a wood vat [called] *toko*.[22] They cover [the vat] with many bamboo mats and put it into a (fermentation) room (*muro*) for about half a day. Around the hour of the Horse they smash the lumps and at that time they add *moyashi* in a ratio of approximately two *go* / (360 cc or ml) to one *koku* (180 l). At the eighth hour of the night, they empty the vat and arrange the rice neatly in the middle of *kōji* drawers (*kōji-futa*), of which they put ten on top of each other. In the course of the following day they turn the rice once, and when evening comes, they spread the rice equally over the whole bottom of the drawer. The drawers being alternatively stacked widthways and lengthways, by the seventh hour of that night [the contents] will have turned into yellow or white *kōji* (Nippon sankai meisanzue, fasc. 1).[23]

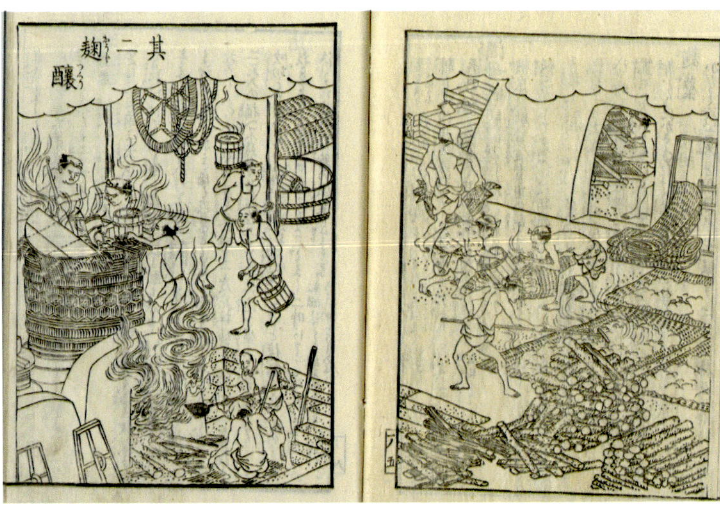

Fig. 5: 'The making of *kōji*', *Nippon sankai meisan zue*, fasc. 1.
Source: Castillon collection. Collège de France.

[22] A *toko* is a container into which cooked rice is put.

[23] This is Wim Boot's translation, with small changes. Chiba, *Nippon sankai meisan meibutsu zue*, pp. 20–21.

Below are some of the key features of the chapter:

Introduction to technical terminology

Honzōgaku was primarily a science of names. It was a common practice for scholars in this field to list the names by which a product was designated. We find this feature in the *Meisan zue*. In all the entries of the book, a significant amount of space is devoted to explaining terminology. In the case of *sake*, the terminology is that used by producers. The author begins with an explanation of the different types of *sake* that can be found on the market: the 'new', 'intermediate', 'early winter' and 'winter' *sake*. In the quotation above, we can note that the technical term *sake kōji* is introduced with a commentary on the Chinese character used to write this term. For Japanese *honzō* specialists who had to contend with the reality of Japan's natural world, it was of utmost importance that each substance was correctly named, i.e. associated with a Chinese character, as this was the best way to prove the antiquity and the prestige of the item. Other technical terms, such as *moyashi*, or *moto*, which are Japanese names, are also introduced with Chinese characters, without any commentary on their origin. Among technical terms, we also find terms such as *hitomoto*, *nakano mi* and *shimai*, which corresponded to the results of the first, second and third series of the addition of rice, *kōji* and water to the *moto*. Such terms belonged to the professional *jargon* that could only be collected on-site. They were evidence that the author or the illustrator had visited the place and had made direct contact with *sake* brewers.

Mention of specialist tools

Similarly, we may note a strong interest in the working equipment throughout *Meisan zue*.[24] In the case of *sake* brewing, the tools that are mentioned are not specific to the field: rakes (*ka'i*), straw mats (*mushiro*), wood vats (*hangiri*), strainers (*koshiki*), mortars (*usu*), and so on. The author stresses the number of tools that are used in a special paragraph entitled 'utensils necessary for brewing *sake*'. The Itami producers used large containers of different volumes because of the need to add water, rice and *kōji* repeatedly to obtain the desired levels of fermentation, and because of the large quantities of *sake* produced. The description mentions the size and the

24 This was already the case in some chapters of the *Meibutsu zue*, such as the entries dealing with mining. See, for example, 'Various mining tools' (*Kanayama shodōgu*) in *Nippon meibutsu zue*, fasc. 1, *f.* 7.v and *f.* 8. Or Chiba, *Nippon sankai meisan meibutsu zue*, p. 200.

number of receptacles to be used at each step. The illustration also emphasizes the way mixtures are poured and transferred from one container to another (see Fig. 6).

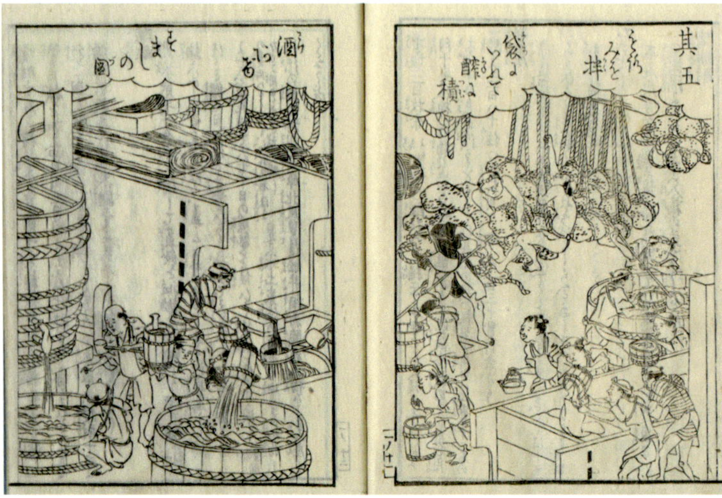

Fig. 6: 'The washing of rice', *Nippon sankai meisan zue*, fasc. 1.
Source: Castillon collection. Collège de France.

Time

Another important element of *sake* brewing was time. The success of brewing depended heavily on control of the fermentation process, which required constant attention. *Sake* could be produced during all seasons, but *sake* brewers in Itami city specialized in winter *sake*. The fermentation process for winter *sake* required more time to complete. It also required more attention. Brewers could spend the night monitoring the smooth running of the operation. The author therefore mentions very precisely the number of days and hours one should devote to each operation or the number of times one should repeat an operation and how often, these things being of utmost importance.

Division of the production procedure into a series of standardized actions

The description presents *sake* brewing as a chain of standardized actions. The author reinforces the impression of standardization by indicating numerical data at each step of the process. In the case of *sake*, the author provides the proportions in which the ingredients should be added in order to obtain the *moyashi*, *kōji*, *moto*, and *moromi*. An example of this can be seen in the previous quotation, which notes, 'Around the hour of the horse they smash the lumps and at that time they add *moyashi* in a ratio of approximately two *gō* to one *koku*.'

The impression of standardization comes also from the fact that each worker was assigned a specific task. The workers represented in the plates are not individualized. They are all dressed the same way (i.e., in a straw loincloth) and act as an essential part of the machinery. In the case of *sake*, it is clear that *Meisan zue* helps to create a representation of production as a standardised meticulously organized chain of action.

3.2. Tuna fishing

We turn now to the example of tuna fishing. As can be seen in the table, fishing appears as a major topic of interest in the *Meisan zue*. During the late Edo period, products such as *katsuo* (skipjack tuna), or *shibi* (tuna) were caught in large quantities throughout the archipelago and dispatched to the nearest cities. Though the *Meisan zue* says nothing about the managerial side of this activity, the description shows that fishing had become a major business by this period, and relied on sophisticated organization and specialized labour. *Meisan zue* notes that 'a fine catch of tuna could reach a quantity of fifty to seventy thousand fish.'[25]

The description, which covers two folios and includes one illustration, proceeds according to the following scheme:

1) Names given to tuna in different parts of Japan
2) Places where tuna is of the highest quality
3) Types of tuna caught according to seasons
4) Ways of selling and eating
5) References found in Chinese or Japanese classics and ancient poems
6) The fishing technique used in Hirado domain[26]
7) An alternative technique used in Wakasa province[27]
8) Description of the tuna and its behaviour
9) Local ways of pronouncing the Chinese character *shibi* 鮪

---

25 Chiba, *Nippon sankai meisan meibutsu zue*, p.100.

26 At this time, Hirado was an important port city in northern Kyūshū where the castle of the Matsura lords was located. Nowadays, Hirado is part of Nagasaki prefecture.

27 Name of an ancient province, located in present-day Fukui prefecture, which is to the north-east of Kyōto.

Meisan zue mentions that tuna fish are caught in many parts of Japan and that they are named differently according to the place and the season in which they are caught.[28] The tuna fishing described here concerns Pacific Bluefin tuna (*Tunnus orientalis*) caught at Hirado. Nowadays this species is known for its huge size (3 meters in length and 450 kg in weight) and its red flesh. It is widely consumed and is known as *kuromaguro* (black *maguro*). The *Meisan zue* mentions *maguro* as a local name used in eastern Japan, i.e. in the Edo region.[29] The author uses *shibi* as the generic name and says that the ones caught between November and mid-January are called *kuro shibi* or black *shibi*. Here we shall look closely at point 6 in the schema above, the fishing technique used in Hirado, a port city located in northern Kyūshū. It offers a good example of a sophisticated technology, which had first been used in Kinki region before reaching the island of Kyūshū by the eighteenth century.[30]

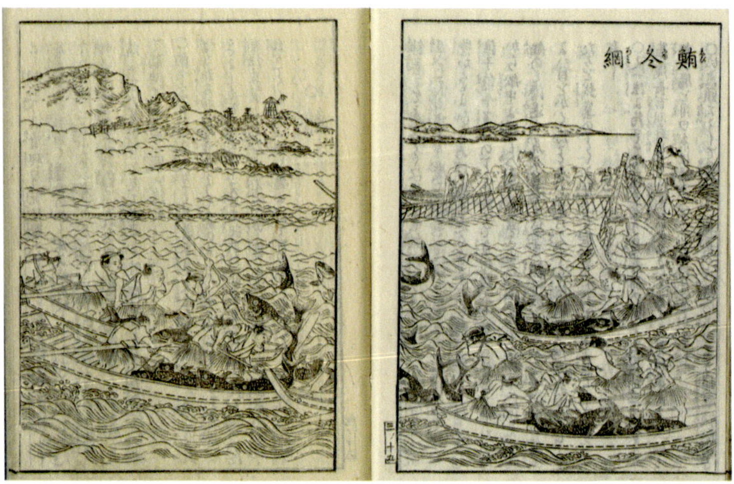

Fig. 7: 'Winter nets for tuna', *Nippon sankai meisan zue*, fasc. 3. Source: Castillon collection. Collège de France.

---

[28] Chiba, *Nippon sankai meisan meibutsu zue*, p. 100.

[29] Chiba, *Nippon sankai meisan meibutsu zue*, p. 99. From the *Meisan zue* we also learn that by this time there was no single designation for this fish in Japan and that *shibi* or *maguro* was much more appreciated in eastern than central Japan, where the *katsuo* was preferred. See Chiba, *Nippon sankai meisan meibutsu zue*, p. 101.

[30] Tajima Yoshiya 'Kinsei kishū ryōhō no tenkai' (Development of the fishing methods of Kii province in early modern period), in Hayama Teisaku (ed.), *Seisan no gijutsu* (Production technology), Nihon no kinsei 4, Tōkyō: Chūō kōronsha, 1992, pp. 211–278.

The Hirado fishing technique is described as follows:

> The fishing net has a mesh of approximately 8 *sun* (24 cm) and a length of 20 *chō* (2 km). It is made of a thin rope. It has a dustpan-like bottom with a bag at its end. The [hauling] ropes [at both ends] are thicker than a thumb and are sunk on the seabed. [see Fig. 7]. Each end is fixed to two boats, and one waits for the fish to gather in large numbers. If the fish come later than usual, the boats can spend two to three months watching the net in vain. There is a watch post on the hill from which schools are located and the number of fish estimated by thousands or tens of thousands. Then, someone shouts *kama-iro kama-iro* (which means 'Get ready') waving a signal flag. At this moment, three small boats called *danbei* start rowing. There are three men in each boat wearing a straw loincloth, a headband, with sleeves tied with a string. The boats come up as fast as the wind. The men seize the bottom of the net and when it is half raised, again the signal flag is waved on the hill. Then, a lot of *danbei* arrive and lift the net together. As the hauling boats approach, the fish spurt out at the surface. When the men hit the fish with rakes or gaffs, the fish wriggle more intensely and jump into the boat on their own. When the fish have finished jumping on board, the fishermen sink the net back on the seabed. Only the boats withdraw [from the site] (*Nippon sankai meisan zue*, fasc. 3). [31]

Here the production process is limited to catching fish, which may seem less elaborate than *sake* brewing at first sight.[32] Yet, one can note features in common with the description of *sake* production seen above. For example, there is a similar focus on:

- varieties and quality of products ("the one caught at Hirado Iwashimizu is of superior quality") [33]
- production area ("large quantities are fished in Munakata in Chikuzen province, Sanuki province, Hirado and Gotō islands"[34])
- production environment, in this case the fishing ground
- equipment (net, boats, rake, hook)

---

[31] Chiba, *Nippon sankai meisan meibutsu zue*, pp. 99–104.

[32] This is not the case for all fish. For example, in the case of bonito (*katsuo*), the description includes the preparation of *katsuo bushi* (*katsuo* flakes), an important ingredient in Japanese food.

[33] Chiba, *Nippon sankai meisan meibutsu zue*, p. 100.

[34] Ibid.

- the sequence of operations
- workers (their clothing and number)
- technical terminology (*danpei*, *kama-iro-kama-iro*)

We also note the complementary role played by the illustration. The illustration gives a global vision of the scene of capture. It shows the fishing ground, which is some distance from the coast, the watch post on the hill, the fishing unit comprising ten small boats with three men aboard each, the large fishing net, the fishermen lifting the net and pushing the fish into boats with gaffs.[35] The picture is faithful to the text and completes it by stressing the size of the fish and the strong commitment of fishermen. Their faces are stern and show a high level of concentration. They look like people engaged in a war against the fish.

Let us return to the fishing technique itself. Though mentioned later in the section, the author establishes a link between the behaviour of the fish and the fishing technique. After a physical description of tuna, the text notes:

> The strength of the fish is located in its head. When the head is turned towards the shore and its tail towards the sea, it is caught more easily because its tail is weak. When, attracted by the heat, it rises to the surface, it is dazzled by the sun and forms benches. The fishermen catch it either to collect the oil or to dry it.[36]

This is a common feature among all of the entries relating to fishing activity. For each type of fish there is specific equipment or a *modus operandi* resulting from the analysis of its behaviour. *Meisan zue* clearly seems to consider that the success of each method comes from this knowledge.

Another striking feature of the description is the focus on the net. Nets are a central subject of the third fascicle, so much so that they appear in the captions of almost all of the plates (see the list below). By this time, fishing nets had become highly sophisticated objects. Their size, structure and material varied according to the type of fish.[37] The author of *Meisan zue* provides accurate dimensions, the category (gillnet, drive-in net, driftnet, etc.),

---

35 For a description of fishing techniques, see Arne Kalland, *Fishing Villages in Tokugawa Japan*, Richmond: Curzon Press, 1995, pp. 99–115.

36 Chiba, *Nippon sankai meisan meibutsu zue*, p. 103.

37 See A. Kalland, *Fishing Villages*, p. 99–115.

and the technical name. He also indicates where the net was set (far or close from the coast, on the seabed, etc.), and the way it was lifted and hauled by fishermen.

Table 2: Plate captions in the fascicule 3 of the *Nippon sankai meisan zue*

Pl. 26 Ise abalones
Pl. 27 Making of abalone ribbons
Pl. 28 Shrimp nets
Pl. 29 Drive-in nets (*oi-ami*) for yellowtail 1
Pl. 30 Drive-in nets for yellowtail 2
Pl. 31 Gillnets for yellowtail
Pl. 32 Winter nets for tuna
Pl. 33 Driftnets for Spanish mackerel
Pl. 34 Nets for flatfish in Wakasa province
Pl. 35 Production of steamed flatfish in Wakasa
Pl. 36 Nets with planchettes for dorado at Emata in Sanuki province
Pl. 37 'Five wisdoms' nets for dorado
Pl. 38 Boats of handliners for mackerel
Pl. 39 Oyster cultivation at Hiroshima

Another focus of the author's interest is the composition of the fishing units, i.e. the number of men and the number of boats involved in the operation. In the case of tuna fishing, the unit was composed of ten boats with three men in each. The illustration highlights the collective and planned nature of the process. As noted for *sake* production, fishermen represented in the figures lack any sign of individuality. They are dressed the same way and each is assigned a specific role in the chain of production (see Fig. 7). This marks a significant difference from *Meibutsu zue*, published forty years earlier, where each worker is illustrated with his own personality. We can therefore say that both the descriptions and the pictures demonstrate that fishery had become a large-scale undertaking, involving a high level of organization, a large number of skilled, knowledgeable fishermen, and sophisticated equipment.

All the products described in the book were 'famous', i.e. known throughout Japan. This meant that these products had been successfully marketed. The book seems to suggest that the success of the products was due to (or at least strongly connected to) the features just mentioned: careful organization of people, precision of gestures, quality of equipment, and a thorough knowledge of natural species.

4. THE INTELLECTUAL CONTEXT OF *MEISAN ZUE'S* PUBLICATION

Now that we have a more accurate view of the content of *Meisan zue*, we may ask why such a book was published and what its underlying purpose was. To answer this, it may be useful to recall the context in which this book was written

Meisan zue was published at a time when *bussangaku* ('product studies) already had a well-established audience of connoisseurs, of which Kimura Kenkadō was a distinguished representative. These connoisseurs, who came from different backgrounds, would hold meetings where they showed each other pieces from their personal collections and exchanged information about curious or rare species. From the 1750s onwards, such exhibitions were held on a yearly basis, in Ōsaka on the initiative of Toda Kyokuzan (1696–1769), or in Edo, under the patronage of Tamura Ransui (1718–1776) and Hiraga Gennai (1728–1780).[38]

Butsurui hinshitsu (Assessment of product quality), published in 1763 by Hiraga Gennai, a major contributor to and organizer of these exhibitions in Edo, can help us identify the species that were of interest to this public and to understand the reasons for their popularity. *Butsurui hinshitsu* is composed of short comments on three hundred species chosen from among the 1,300 items that were exhibited at the fifth and final exhibition that Hiraga Gennai organized in Edo. As the name *yakuhin-kai* or "medicinal articles gathering" given to these exhibitions indicates, *materia medica* was a major concern for these scholars, who considered it their priority to reduce Japan's dependence on foreign imports (although rare or exotic species were also a major attraction at these exhibitions). *Butsurui hinshitsu* adopted the same classification as *Bencao gangmu*: metals, stones, herbs,

[38] For more information about these exhibitions, see Federico Marcon, *The Knowledge of Nature and the Nature of Knowledge in Early Modern Japan*, Chicago: The University of Chicago Press, 2015, pp. 207–227.

grains, vegetables, fruits, trees, insects, scaly creatures, quadrupeds. Well-known species, such as cinnabar, mercury, amber, Chinese licorice (Jp. *kanzō*), ginseng, are listed alongside exotic species such as the Asian house shrew (*jakō nezumi*), which was only found in Nagasaki, sugar cane imported by Satsuma domain from the Ryūkyū islands, or species that were known only by their Dutch names. The logic behind the selection is not immediately apparent. Nevertheless, three interconnected intentions can be identified: 1) to disseminate knowledge about unknown products; 2) to inform how imported products could be grown or produced in Japan; and 3) to estimate and compare the quality of the products on the market, according to their origin.[39]

This orientation is particularly explicit in the sixth fascicle of the *Butsurui hinshitsu*, entirely devoted to describing the production methods of ginseng and sugar cane. Hiraga Gennai explains in a short introduction that, despite the efforts of shogunal authorities to collect seeds from China, Korea and European countries, and though dozens of seeds were preserved in official gardens, it could hardly be said that they were widely disseminated. He then announces the aim of his short study:

> If these seeds could be planted and produced everywhere in our provinces, *the profit would be great.* In particular, while ginseng[40] and sugar cane are widely consumed, as we do not know how to cultivate them, and planting them doesn't give much result. I have gathered here, in the public interest, the main points of what I have learnt from my own experience and from the methods used by many others.[41]

From this quotation we can infer that for *honzōgaku* connoisseurs like Hiraga Gennai, these exhibitions served the important objective of disseminating information about production methods in order to encourage and facilitate domestication. It is thus clear that Hiraga's aim was not only scientific but also political. He is known

[39] Concerning *Butsurui hinshitsu*, see for example Jōfuku Isamu, *Hiraga Gennai no kenkyū* (Research on Hiraga Gennai), Ōsaka: Sōgensha, 1976, pp. 111–134.

[40] Regarding Shogun Yoshimune's policy for developing ginseng production in Japan, see Yabe Ichirō, *Edo no honzō*, pp. 84–90.

[41] Hiraga Gennai, *Butsurui hinshitsu* (*Assessment of product quality*), fascicle 6, supplement (*furoku*), Hiraga Gennai sensei kenshōkai (ed.), *Hiraga Gennai zenshū* (Complete works of Hiraga Gennai), vol. jō, Tōkyō: Hiraga Gennai sensei kenshōkai, 1932, p. 157.

to have been one of those who, at this time, worked actively and consciously to support the mercantile policy of the bakufu, under the government of Tanuma Okitsugu (1719–1788).

It is important, however, not to overvalue Hiraga Gennai's contribution. He was not the first to highlight production issues in books. *Honzō* books, especially *Bencao gangmu*, have never been devoid of reflections on production methods. However, the books that addressed the issue most directly were agricultural treatises. The purpose of agricultural treatises (*nōsho*) was to provide farmers with general knowledge about agriculture and to introduce them to new methods, especially those in use in more advanced provinces. Hence *Nōgyō zensho* (*Compendium of agriculture*) in 10 fascicles by Miyazaki Yasusada (1623–1697), the most complete and influential book of its kind published during Edo period, provides the standard method of cultivation for more than a hundred specimens of plants, grains and trees, with a number of practical tips. The book, first published in 1697, also contains illustrations, though not as many as the *Meisan zue*. It was reprinted several times, especially in the 1780s, and can be considered to be widely known in the late eighteenth century.[42] Though *Nōgyō zensho* was written before Japan entered the era of commercial agriculture in the second half of the Tokugawa period, it provided information on a wide range of topics that could be put to use by farmers seeking to develop their land. The interest in domesticating plants of foreign origin is not absent, as can be seen in Miyazaki's remark about sugar cane: "If this technique [of planting sugar cane] were to be mastered and if we were able to produce it here at home, it would be a way of preventing our wealth from being absorbed by foreign countries."[43] It is interesting to note that though the book is mainly dedicated to methods of cultivation (choice of soil and seeds, suitable times to sow, plant, harvest, how to fertilize the soil, etc.), it also contains the manufacturing procedures for major products, such as tea and tobacco.[44]

42 For an overview of the *nōsho* genre, see Tsukuba Hisaharu, *Nihon no nōsho: nōgyō wa naze kinsei ni hatten shita ka* (Agricultural treatises in Japan: Why agriculture developed in early modern period?), Tōkyō: Chūō kōron sha, 1987.

43 Miyazaki Yasusada, Kamiya Takao (annotated by), *Nōgyō zensho* (Compendium of agriculture), Tōkyō: Iwanami shoten, 1936 (first printed), p. 204.

44 Regarding tea production, see Guillaume Hurpeau, *Histoire du thé au Japon: Techniques culturales et de fabrication du thé à l'époque Edo*, PhD Thesis, Ecole Pratique des Hautes Etudes, Paris, 2018.

However, there are also notable differences between a book like *Nōgyō zensho* and our *Meisan zue*. The former was written by a man who practiced agriculture and who addressed it to his peers. This is not the case for the *Meisan zue*, whose author (or authors) clearly belonged to the literate circles of Ōsaka and was not involved in the production activities describes in the book. Nor was the *Meisan zue* intended for workers or people involved in such activities. This is made evident by the total absence of any didactic style, such as that which characterizes Miyazaki's book. With these considerations in mind, what could have motivated such a work?

With Hiraga Gennai, we saw that scholars could be driven by political considerations. They could seek to influence domainal or village policy directly. Indeed, by providing a detailed description of production techniques, and showing the success obtained in other provinces, the book might have showed a path to those who held positions of governance in the country.

If these remarks shed light on the underlying intentions of this book, they do not fully explain why such a systematic and detailed description of production processes, in particular for manufactured products, appeared at this time in the *Meisan zue*. One possible source of inspiration could be Song Yinxing's (1587–1666?) *Tiangong kaiwu* (Works of heaven and the inception of things), a late Ming opus printed in China in 1637.[45] The book comprises 3 volumes, divided into 18 sections. It covers all the main branches of production in late Ming China: 1) Varieties of grains, 2) Clothing material 3) Dyeing 4) Milling 5) Salt production 6) Sugar production 7) Porcelain production 8) Casting 9) Ships and Vehicles 10) Forging 11) Calcination of stones 12) Oil production 13) Papermaking 14) Varieties of Metals 15) Weapons 16) Vermilion and ink 17) Fermentation 18) Pearls and gems. As can be seen from its table of contents, the *Tiangong kaiwu* has an encyclopaedic orientation. It devotes substantial space to the description of production techniques, equipment and labourers' work.

Song Yingxing was a late Ming scholar who, experiencing the chaos of the Ming collapse and the failures of Ming orthodoxy, advocated a return to true knowledge, rooted in the mundane world and in common sense.[46] We do not know the extent to

---

45 Translation by Dagmar Schäfer. See Dagmar Schäfer, *The Crafting of the 10,000 Things, Knowledge and Technology in Seventeenth-Century China*, Chicago: The University of Chicago Press, 2011.

46 Schäfer explains Song's interest in crafts by his epistemological stance: '[Song] saw the revelation of universal principles within the performance of crafts and

which Japanese scholars were sensitive to Song's rhetoric and ideals. However, Song's rich writing on agricultural and non-agricultural production techniques certainly caught their attention, as evidenced by the many points of convergence between his work and both *Meisan zue* and *Meibutsu zue*.

Tiangong kaiwu, first printed in 1637, did not receive wide recognition in China, but the book was never completely forgotten and survived until modern times because of its highly informative content.[47] In Japan, while it is mentioned only occasionally in late-seventeenth century sources, its influence suddenly grew from the mid-eighteenth century onward, leading to its reissue in 1771. Interestingly, the Japanese edition was printed by an Ōsaka publisher and was based on a manuscript belonging to Kimura Kenkadō, the author of *Meisan zue's* preface. This confirms the strong interest that Ōsaka publishers and scholars had in the issue of production techniques in the mid-eighteenth century.

As for the reception of the book, we know that *Tiangong kaiwu* was frequently quoted in *Butsurui hinshitsu*, which also contains a slightly modified illustration from the *Tiangong kaiwu*, in the sixth fascicle (see Fig. 8 and Fig. 9). *Meisan zue* also refers explicitly and implicitly to *Tiangong kaiwu* in the sections devoted to porcelain and lime production, two important topics in the Chinese compendium. As for the illustrations, while similarities can be observed between those of *Meibutsu zue* and *Tiangong kaiwu*, the same cannot be said for *Meisan zue*, whose representation techniques seem to have other sources of inspiration.

There is however one important topic in the *Meisan zue* that can be found neither in the *Tiangong kaiwu* nor in the agricultural treatises. These are seafood products, the sources for which seem to have been Japanese books of the *honzō* tradition.[48]

---

technology; it manifested a cosmological order that man had to comprehend in order to be able to quell the chaos that ruled his era.' Schäfer, *The Crafting of the 10000 Things*, p. 17. Schäfer also emphasizes Song's profile as a natural philosopher: 'In his attempt to provide a universal all-encompassing theory of *qi*, Song relied on existing theory, expanding it when it proved insufficient to explain the phenomena he so closely observed.' (Ibid, p. 233).

47 Concerning the history of the book, see Schäfer, *The Crafting of the 10,000 Things*, pp. 259–262 and for its editions, pp. 274–275.

48 The *Honchō shokkan* and *Yamato honzō* are abundantly quoted.

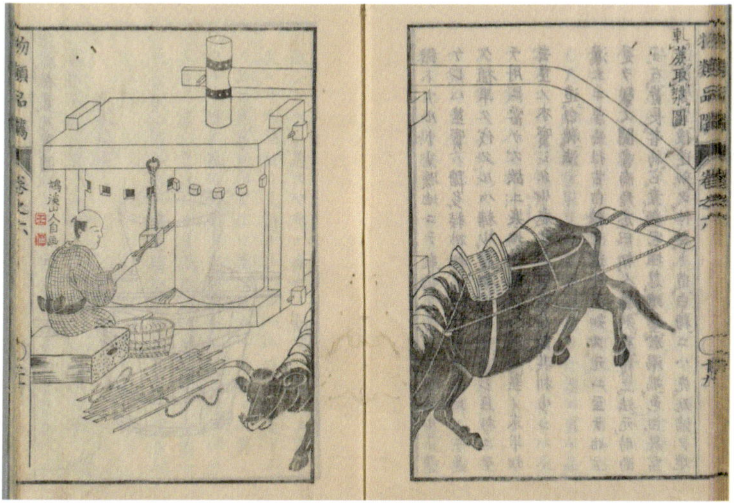

Fig. 8: *Butsurui hinshitsu* (1763), fasc. 6.
Source: National Diet Library Japan, ID 000007315267 http://dl.ndl.go.jp/info:ndljp/pid/2555270?tocOpened=1 (image 21)

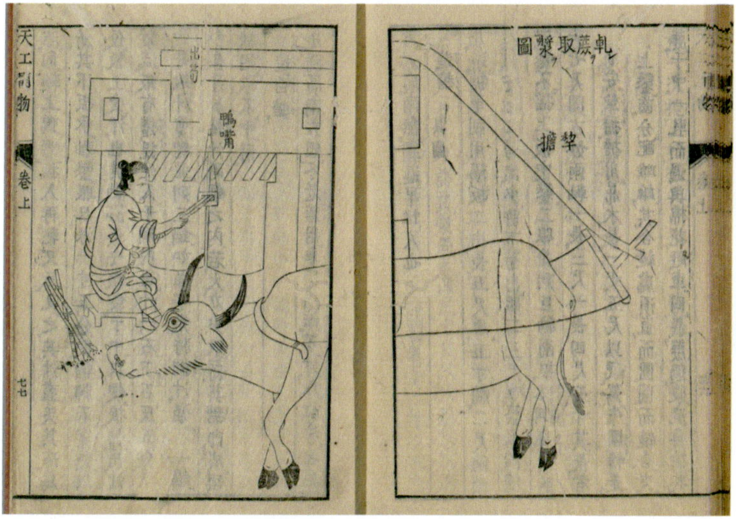

Fig. 9: *Tiankong kaiwu* (Japanese edition, 1771), fasc. 3.
Source: National Diet Library Japan, ID. 000007598449 http://dl.ndl.go.jp/info:ndljp/pid/2556157?tocOpened=1 (image 31)

5. CONCLUSION

We have seen in this paper that descriptions of production techniques became a major focus for Japanese scholars from the mid-eighteenth century, and that Ōsaka publishers played a significant role in this development. In this respect, *Nippon sankai meisan zue* is, with its detailed and systematic descriptions based on direct observation, and with its wide range of subjects, a landmark publication.

The book sits at the intersection of several editorial genres. These include *honzō* or *bussan* books, which were in full expansion after the exhibitions organized in Ōsaka and Edo, and agricultural treatises, which also enjoyed great success, though printed books read nationwide like the *Nōgyō zensho* were exceptional. This genre tended to merge with *bussangaku* books from the nineteenth century onward, as shown by works of Ōkura Nagatsune (1768–?).[49] We have also seen that *Meisan zue* shared some features with *Tiangong kaiwu*, reprinted in Ōsaka in 1771. This encyclopaedic book covering the main fields of production of the Chinese empire was not only an invaluable source of information, it also inspired Japanese scholars, who discovered that production techniques could be a topic of study for a prominent Chinese scholar.[50]

This convergence of editorial genres illustrated by *Meisan zue* was motivated by the concern of scholars, and more particularly of Ōsaka scholars, to contribute to the enrichment of the country. If the initiative stemmed from scholars and publishers, these books were aimed at a audience that cannot be easily characterized. At the end of the eighteenth century, the readership of scholarly books expanded considerably; if Hiraga Gennai was addressing the powerful people who could finance his projects, there were also among his readers merchants in search of good deals, village headmen seeking to pay less taxes, poor samurai seeking to make a fortune, or lords dreaming of restoring the finances of their domain.

---

49 See, for example, his *Kōeki kokusankō* (Thoughts on the broad benefits of national production, 1843) which can be categorized as a work of *bussangaku*.

50 These effects are visible in the works of authors like Satō Nobuhiro (1769–1850), who takes up the notion of '*kaibutsu*' (ch. *kaiwu*), 'inception of things', in his work *Keizai yōryaku* (Essentials of economic policy). See Bitō Masahide, Shimazaki Takao ed., *Andō Shōeki, Satō Nobuhiro*, Nihon shisō taikei 45, Tōkyō: Iwanami shoten, 1977, p. 534.

Table 1: Abridged Table of Contents of the *Nippon sankai meisan zue*

Volume 1 (1 entry; 12 sub-entries)

- *Sake* brewing

Volume 2 (11 entries; 20 sub-entries)

- Varieties of stones
- Varieties of mushrooms
- Bees and beeswax
- Varieties of prey (falcon, duck, bear)

Volume 3 (11 entries; 1 sub-entry)

- Fishing products :
- Abalone, pearls, shrimp, yellowtail, tuna, mackerel, flatfish, small sea bream, oyster etc.

Volume 4 (10 entries; 2 sub-entries)

- Fishing products :
- skipjack tuna, urchin, sea cucumber, *shirouo*, hard clams, *gori*, trout, lamprey, octopus, *kajika*

Volume 5 (8 entries)

- Jellyfish
- Lime
- Ceramics
- Echigo fabrics
- Sea lions
- *Kombu*
- Foreign products
- Dutch ships

REFERENCES

Bitō Masahide, Shimazaki Takao (eds), *Andō Shōeki, Satō Nobuhiro*, Nihon shisō taikei 45, Tōkyō: Iwanami shoten, 1977.

Chiba Tokuji (ed.), *Nippon sankai meisan meibutsu zue*, Tōkyō: Shakai shisō-sha, 1970.

Higuchi Hideo, '*Kaisetsu*' (Introduction), in the facsimile edition of the book *Nippon sankai meisan zue*, Tōkyō: Meicho kankō kai, 1979, p. 1–11.

Hiraga Gennai, *Butsurui hinshitsu* (*Assessment of product quality*), fascicle 6, supplement (*furoku*), Hiraga Gennai sensei kenshōkai (ed.), *Hiraga*

Gennai zenshū (Complete Works of Hiraga Gennai), vol. jō, Tōkyō: Hiraga Gennai sensei kenshōkai, 1932.

Hurpeau, Guillaume, *Histoire du thé au Japon: techniques culturales et de fabrication du thé à l'époque Edo*, PhD Thesis, Ecole Pratique des Hautes Etudes, Paris, 2018.

Jōfuku Isamu, *Hiraga Gennai no kenkyū* (Research on Hiraga Gennai), ōsaka: Sōgen-sha, 1976.

Kalland, Arne, *Fishing Villages in Tokugawa Japan*, Richmond: Curzon Press, 1995.

Marcon, Federico, *The Knowledge of Nature and the Nature of Knowledge in Early Modern Japan*, Chicago: The University of Chicago Press, 2015.

Métailié Georges, 'The *Bencao gangmu* of Li Shizhen: An Innovation in Natural History?', in Elisabeth Hsu (ed.), *Innovation in Chinese Medicine*, Cambridge: Cambridge University Press, 2001, pp. 223–224.

Miyazaki Yasusada, *Nōgyō zensho* (Compendium of Agriculture), first printed version: Tōkyō: Iwanami shoten, 1936.

Ōsaka rekishi hakubutsukan (eds), *Kimura Kenkadō: Naniwa chi no kyojin* (Kimura Kenkadō, a giant of ōsaka's learning), Kyōto: Shibunkaku shuppan, 2003.

Schäfer, Dagmar, *The Crafting of the 10 000 Things, Knowledge and Technology in Seventeenth-Century China*, Chicago: The University of Chicago Press, 2011.

Shinpo Hiroshi and Hasegawa Akira, 'Shōhin seisan ryūtsū no dynamics' (Dynamics of the production and circulation of commercial goods), in Hayami Akira and Miyamoto Mataji (eds), *Keizai shakai no seiritsu 17-18 seiki* (Development of economy and society in the seventeenth to the eighteenth century) (Nihon Keizai 1), Tōkyō: Iwanami shoten, 1988, pp. 217–270.

Shirahata Yōzaburō, 'The Printing of Illustrated Travelogues in eighteenth century Japan', in Susanne Formanek & Sepp Linhart (eds.), *Written texts – Visual Texts Woodblock-printed Mass Media in Early Modern Japan*, Leiden: Hotei Publishing, 2005, pp. 199–214.

Tajima Yoshiya, 'Kinsei kishū ryōhō no tenkai' (Development of the fishing methods of Kii province in early modern period), in Hayama Teisaku (ed.), *Seisan no gijutsu* (Production Technology), Nihon no kinsei 4, Tōkyō: Chūō kōronsha, 1992, pp. 211–278.

Tsukuba Hisaharu, *Nihon no nōsho: nōgyō wa naze kinsei ni hatten shita ka* (Agricultural Treatises in Japan: Why Agriculture Developed in Early Modern Period?), Tōkyō: Chūō kōronsha, 1987.

Yabe Ichirō, *Edo no honzō: yakubutsugaku to hakubutsugaku* (The *materia medica* of Edo: from Pharmacopoeia to Natural History), Tōkyō: Saiensu-sha, 1984.

Yunoki Manabu, *Sakezukuri no rekishi* (History of *sake* Making), Tōkyō: Yūsankaku, 2018 (first printed in 1975 as *Nihonshu no rekishi*).

Internet source:
'Sake'. Britannica Academic, *Encyclopædia Britannica*, 4 Jan. 2019. (academic-eb-com.rproxy.sc.univ-paris-diderot.fr/levels/collegiate/article/sake/65000. Accessed 22 Mar. 2019).

2

Vehicles of Knowledge: Japanese Technical Drawings in the Pre-modern Era, 1600–1868

Erich PAUER

1. INTRODUCTION

People transfer knowledge in various ways. The simplest form of transmission is to speak or to hear, but such knowledge is not sustainable and is easily lost. One way to make knowledge sustainable is to write it down. Once in a written form – this can be a text, but also pictures and drawings – it then can be forwarded. Therefore, for us, books, technical drawings, and the like seem to be the most suitable 'vehicles' of knowledge. However, not every drawing is a vehicle of knowledge, particularly when it comes to technical knowledge. One must examine the intrinsic link between image, technology, and the rules set for codification of practical knowledge in printed form. Different kinds of such prescriptive literature and so-called 'how-to books', to use a contemporary term, such as books of secrets and recipes, show how practices became formalized and codified.

2. SIMPLE DESCRIPTIONS OR TECHNICAL MANUALS?

Although writing had already arrived from China by the fourth century CE, the Japanese book tradition dates back only to the eighth century. These Japanese writings are usually not books with technical content, as they contain myths, representations of history, and geographical writings. However, some refer to buildings, roads, watchtowers, canals, dams, ponds and trenches, and

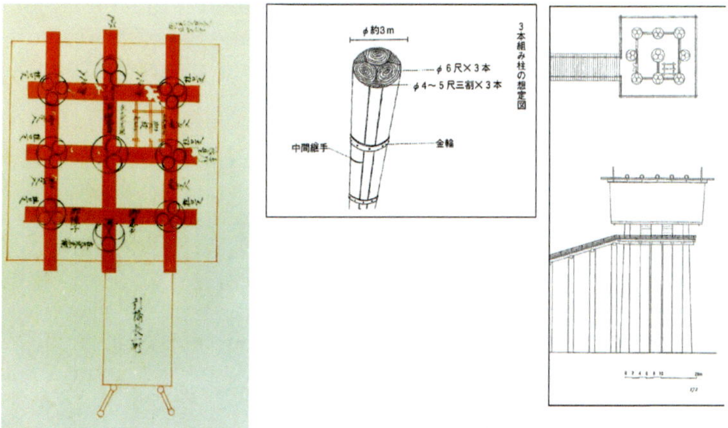

Fig. 1: Instruction chart for the base of the Izumo Shrine: Left the basis with nine pillars; each individual pillar is made up of three tree trunks (figure in the middle); Right: The image of the Shrine on high pillars.
Source:Ōbayashi-gumi purojekuto chimu (publ.), *Yomigaeru kodai kensetsu jidai*, Tōkyō: Ōbayashi-gumi 2002, pp. 269 and 272; Ōbayashi-gumi (eds), *Kodai Izumo-taisha no fukugen*, Tōkyō: Gakusei-sha 2000, p. 172.

especially temples and shrines. These provide evidence of a differentiated class of craftsmen with a marked hierarchy. It can be assumed that this hierarchy was based on skill. Unfortunately, these books give us no details about how these different craftsmen were trained, and contemporary technical textbooks or manuals for these trades do not exist.

Aside from references to craftsmen in ancient writings, pictorial representations of craftsmen at work, or technical details relating to the construction of buildings (mostly temples or shrines), are rare. However, a strange object stands out in ancient writings: it is the first evidence of a 'technical drawing', or the passing on of technical knowledge in written form.

The illustration shows the base of the Shintō shrine at Izumo, one of the oldest religious sites in Japan, near the Sea of Japan. What we see is a cross-sectional view showing the shrine's pillars, each formed of three tree trunks. The drawing, which is obviously 'technical', dates back to the eighth century and is thus the oldest drawing of its kind in Japan. The illustration on the far right, which is based on the results of an archaeological excavation, shows a reconstruction of the entire shrine, with the floor plan above, and a side view. The shrine rested on these high pillars. This is a simple technical drawing showing the basic structure of a building of unknown shape held by some pillars. Archaeologists assume that this is the base of a large Shintō shrine standing on high pillars.[1]

At first sight this drawing is only a single sheet of paper and does not pass on knowledge about the entire construction of the shrine, which would have enabled someone to erect a similar building elsewhere. If, however, one takes a closer look, this rare original chart tells us that a certain technical understanding about 'reading' such a cross-section drawing must have existed. Such drawings are not self-explanatory, but must be decoded with a previously acquired knowledge of how to create such drawings. Given the high-level craftsmanship of the time, which is described in detail in contemporary literature, the draft may have been used to pass on certain ideas about details on building construction.

It is not clear whether this kind of document, of which this is the sole example, belongs to a traditional Japanese building technique, i.e. a technique developed in Japan itself. As there are no similar charts found shortly before or after the probable time at which this document was produced, one can speculate that it represents a foreign (probably Chinese or Korean) technique brought to Japan by foreign craftsmen. The lack of similar documents from later centuries also suggests that this document was used only in the Izumo area and did not trigger the use of similar paper plans elsewhere in Japan. One can also assume that the knowledge was passed on to others orally and that further plans on paper were therefore not necessary.

The following centuries were marked by a courtly culture. Courtiers were not usually interested in actively engaging in what was considered a relatively dirty craft, but were interested in the results of the work of the craftsmen – and such results were often shown on picture scrolls or folding screens. It is there that we can also find representations of various crafts. In such drawings, certain activities of craftsmen, their tools, and their application are represented pictorially. As such artistically valuable

1 Ōbayashi-gumi purojekuto chīmu (publ.), *Yomigaeru kodai dai kensetsu jidai – kyodai kenzōbutsu wo fukugen suru* (Reviving the ancient period of large-scale architecture – reconstructing giant buildings), Tōkyō: Ōbayashi-gumi, 2002; Ōbayashi-gumi (ed.), *Kodai Izumo-taisha no fukugen* (The reconstruction of the ancient Izumo shrine), Tōkyō: Gakusei-sha 2000.

screens were part of the interior of upper-class living quarters to which ordinary craftsmen usually had no access, the images could not have served as carriers of knowledge for craftsmen about craft techniques or the processing of certain materials. Moreover, as only certain operations, devices, simple machines, and certain activities are shown, such illustrations on folding screens cannot be understood as textbooks or teaching material. Although these screens are not instructions for the construction of buildings or other technical items, we can at least determine some of the equipment that craftsmen used at the time, and infer from it what skills they possessed.

However, the situation had considerably changed a few centuries later. By the end of the twelfth century, a warrior class, whose power base was land ownership, had evolved. For their own protection, but also to exercise power, they built castles. At first, these were surrounded by a simple clay barrier and later, starting in the seventeenth century, by high and curved walls of stone. Building these increasingly bold castles presented a challenge to craftsmen.

In contemporary sources, names of the architects of such castles are often given along with instructions on where and how to build them. Detailed floor and ground plans, as well as sophisticated and detailed instructions with drawings illustrating certain construction techniques, became increasingly common. Such manuals were used as templates for the heads of the various groups of artisans and could be considered as a first technical guide for improving the practices of artisans and as early examples of a written transmission of knowledge.

Craftsmen thus had to be educated, not only in physical skills, but also in the ability to 'read' drawing specifications. A first kind of transfer of knowledge by special teachers (architects, warriors) arose here.

However, to stay with this example, how houses were built, how iron hoe shoes were manufactured, or how ploughshares were forged was not communicated by drawings. For this purpose, 'tacit knowledge' or 'implicit knowledge' is needed due to a lack of corresponding explanation. This knowledge is embodied in 'practical ability', which is not explicitly presented and passed on in words or in written form because the corresponding possibilities, such as a book culture, writing culture, and duplication are still lacking.

Unfortunately, hardly any documents have survived that could be described as 'vehicles of knowledge', i.e. documents that provide detailed technical information and that were intended to be passed on for guidance. Only rarely, rough technical sketches, regarding individual details appear in connection with the construction of castles.

3. KNOWLEDGE TRANSFER FROM THE CONTINENT: EXAMPLES OF AGRICULTURE

During the period of peace following Japan's unification in 1600, economic consolidation and population growth in the cities resulted in a big rise in food demand. Meeting this demand became a major concern. Feudal lords therefore became interested in how to increase food production and, correspondingly, supported agricultural development.

The shift to small-scale land cultivation after 1600 led to changes in the importance of certain tools. The intensification of agriculture and an increase in the total land area used for agriculture made new irrigation equipment and irrigation facilities necessary. Furthermore, increasing commercialization of agriculture resulted in an increase in the number of plants grown. Therefore, new methods of cultivation, as well as new techniques for processing the foodstuffs produced, were needed.

As in previous centuries, Chinese sources played an important role at first. A famous treatise on agriculture, *Nung-chêng ch'uan-shu* (Collected writings on agriculture), was imported to Japan shortly after its first printing in China in 1639. This work summarized much existing knowledge of Chinese agriculture and illustrated the importance of experimentation to promote agricultural knowledge. As a result, new crops, grasses, and trees, as well as farming tools that were previously unknown to the Japanese, spread in use throughout the seventeenth century. Moreover, this book triggered a wider interest in agriculture and, only a few decades later, Japanese agronomists began to publish works, modelled on this and other Chinese writings, detailing how to improve agricultural output. These Japanese agricultural treatises, called *nōsho*, are a genre, that offers mainly practical and artisanal knowledge based on experience. As there were a certain number of people with reading and writing skills even in rural areas, one can assume that the knowledge transmitted in the books reached its intended audience. The authors were educated farmers rather than intellectuals, who either focused on certain regions or included examples from across the country, using text and images (i.e. pictorial drawings) to promote their agendas in support of their proposals.[2] Improvements in printing technology and a wider range of distribution channels in the seventeenth century resulted in the circulation of several such publications.

To function as a vehicle of knowledge, these writings were mostly written in a simple style and used the Japanese syllabary with only a few Chinese characters so that a literate, but not highly educated, agricultural population could read them.

Two developments can be observed in this genre. One is the evolution of books in general, which increasingly covered various features of agricultural techniques, sowing, soil, fertilizers, pest control, and others. At first, these manuals were often only handwritten and circulated in a limited number, and cannot be regarded as 'how-to books' in a general sense, as they simply described techniques the writer used in his demarcated living environment. Only later, when books offered a variety of different possibilities or examples (e.g. of devices), farmers could choose the one that made the most sense or seemed the most useful, by comparing their own situation with the others.[3] The other is the advancement of images developing from mere illustration to a kind of technical drawing.

Not only tools of Chinese origin but also new inventions native to Japan were introduced through such treatises.

An important agricultural treatise with a nationwide focus is the *Nōgyō zensho* (Compendium on agriculture) by Miyazaki Yasusada,

[2] For an introduction to this genre of books, see Iinuma Jirō, *Nihon no nōsho* (Japanese agricultural treatises) (Chūkō shinsho 852), Tōkyō: Chūō-kōron-sha, 1987; also Iinuma Jirō, *Kinsei nōsho ni manabu* (Learning from the agricultural treatises of the early modern period) (NHK bukkusu 271), Tōkyō: Nihon hōsō shuppan kyōkai, 1976; for the treatises cited in this article see also *Furushima Toshio chosaku-shū*, Tōkyō: Tōkyō daigaku shuppan-kai, 1975, vol. 5, especially the chapter on agricultural treatises in the second half of the seventeenth century on pp. 263–470; see also Erich Pauer, *Technik-Wirtschaft-Gesellschaft. Der Einfluss wirtschaftlicher und gesellschaftlicher Veraenderungen auf die Entwicklung der landwirtschaftlichen Geraete in der vorindustriellen Epoche Japans ab dem 17. Jahrhundert*, (Beiträge zur Japanologie 4), Wien, 1973. See also Jennifer Robertson, 'Japanese Farm Manuals: A Literature of Discovery', in *Peasant Studies*, vol. 11, no. 3 (Spring 1984), pp. 169–194.

[3] This comes close to a situation which Joel Mokyr would have called a 'market for ideas'; see Joel Mokyr, *A Culture of Growth: The Origins of Modern Economy*, Princeton: Princeton UP 2018.

published in 1697 and modelled on the Chinese *Nung-chêng ch'uan-shu* (Compendium on agriculture) published in 1639.[4] That the explanations of the different kinds of cash crops and their treatment, various trees, and planting techniques were the main parts of this publication indicates its focus on the improvement of agriculture.[5] It also provided practical instructions. It extols the benefits of certain new devices and recommends their imitation. The focus of this treatise, however, was on the promotion of commercial agriculture, while the technical dimension and the transfer of technical knowledge required to copy the tools and devices played a minor role.

Consequently, the illustrations given at the beginning of volume 1, which do not correspond to the later text, are not very helpful for understanding how the various devices are built, and even their use is not well documented. Therefore, the book is neither a textbook nor a manual for the agricultural technique in a narrow 'technical' sense concerning the use and making of farming tools.

Moreover, the illustrations have been cut by a wood block carver who obviously had little understanding of agricultural tools and their use. For instance, the tub for scooping water does not function in the way shown. Other illustrations (e.g. the hoe, plough, and mill) are very unclear. The illustrations may have been mere tools to further the interest of the target readership, namely the educated and wealthy farmers.

---

4 The *Nōgyō zensho* was reprinted several times during the Edo period (1600–1868) and also received attention in the modern era. There is an edition of the *Nōgyō zensho* in three volumes (the text transcribed in modern Japanese for easy reading but lacking the plates) published in Tōkyō 1941 in a series titled *Nihon hyakushō bunko* (Library of Japanese farmers) vols. 2, 3 and 4. In recent years, the *Nōgyō zensho was* included in a collection of agricultural treatises of the Edo period (with annotations): Nōsan gyoson bunka kyōkai (ed.), *Nihon nōsho zenshū* (A collection of Japanese agricultural treatises), vols. 12 & 13, Tōkyō, 1978.

5 The volumes contain the following contents:

1 General agriculture; Text and frontispiece (Text by Kaibara Ekiken)
2 The five grains (wheat, rice, beans, millet: *awa* nd *kibi*; 99 kinds
3 Vegetables; 96 kinds
4 Vegetables; 923 kinds
5 Wild grasses; 98 kinds
6 *Sansō* (Three important grasses: Hemp, indigo and safflower); 91 kinds
7 *Shiboku* (Four important trees: Mulberry, lily, lacquer and tea); 94 kinds
8 Fruit trees; 97 kinds
9 various trees; 95 kinds
10 Culture (Livestock and poultry); 93 kinds, Medical plants; 922 kinds
11 Appendix

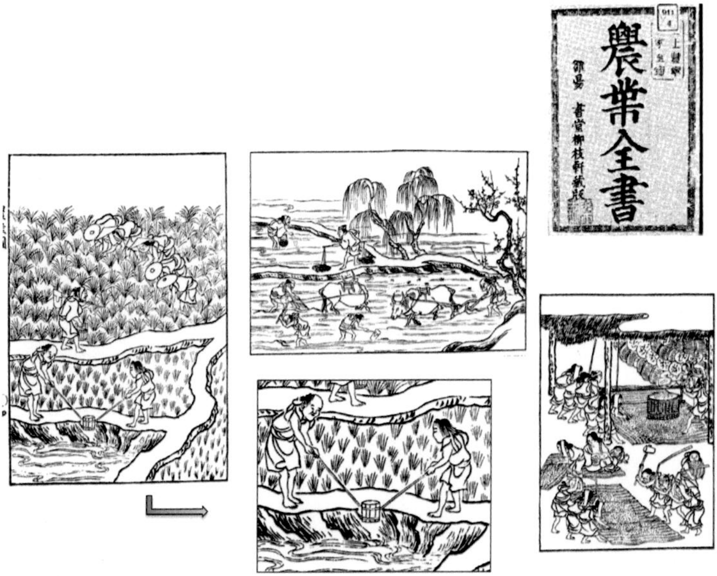

Fig.2: Title page and illustrations (ploughing, scooping) from the *Nōgyō zensho* (orig. 1697); a) ploughing b) irrigation c) irrigation (detail) d) threshing and grinding. The plates are taken from a reprint of the *Nōgyō zensho* from the Tenmei period (1781–1789) in the author's collection. See also Nōsan gyoson bunka kyōkai (eds), *Nihon nōsho zenshō* (A collection of Japanese agricultural treatises), vol. 12, Tōkyō, 1978, pp. 35–43.

However, unlike the *Nōgyō zensho*, which focused on a readership of more educated and wealthy farmers, from the 1680s onward, several agricultural treatises with a regional focus, appeared. In such treatises, we find many – often more accurate – drawings of agricultural tools.

An early example of such a regional treatise is the *Aizu nōsho* (Agricultural treatise of Aizu) from 1684, which was aimed at farmers in the region of Aizu in northern Japan, after which the treatise its named. Based on many years of observation, the author analyses the relationship between the type of soil, fertilizer, rice, and season. He also indicates peculiarities of ploughing, weeding, and furrows and gives advice on special sowing methods.[6]

6 For details on the *Aizu nōsho* see Ono Takeo, *Aizu nōsho*, Tōkyō: Itō Shoten, 1944 (with extensive supplementary annotations) and a more recent reprint in Nōsan gyoson bunka kyōkai (ed.), *Nihon nōsho zenshū* (A collection of Japanese agricultural treatises), vol. 19, Tōkyō, 1982.

As this is somewhat like a 'how-to book', it contains only general information. Others, such as a treatise titled *Hyakushō denki* (The life of a farmer), which was probably written in the 1680s, or a book titled *Kōka shunjū* (Farming from spring to fall), contain much more detailed drawings of agricultural equipment, which aimed explicitly at encouraging the reproduction of farming tools.

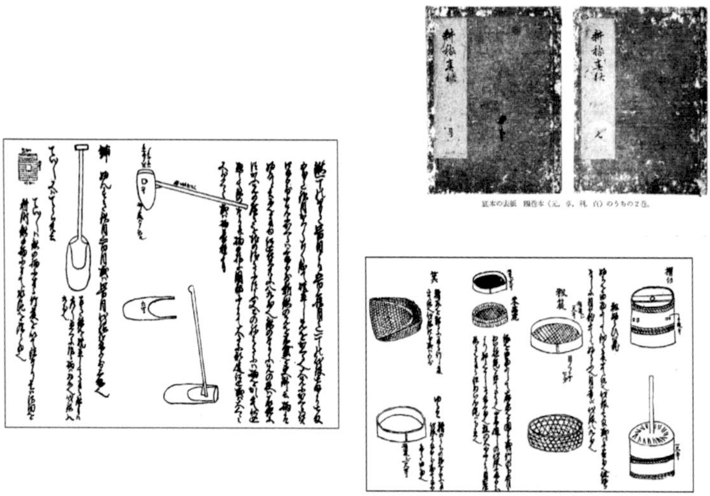

Fig. 3a

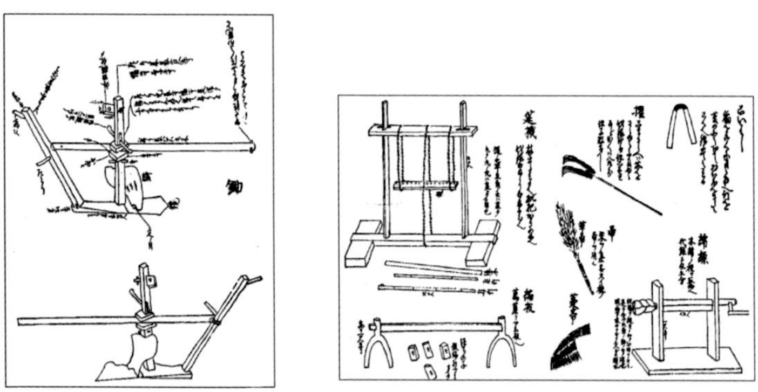

Fig. 3a and 3b: Drawings from the *Kōka shunjū* (1707 ?). Above: various hoes, a spade, tools for winnowing, sieves and grinders; below: a plough, tools for threshing and making sacks for the rice. Source: NKDT (Nihon keizai daiten) 1928 vol. 21, pp. 361, 363, 375, 377; See also Nōsan gyoson bunka kyōkai (eds), *Nihon nōsho zenshō* (A collection of Japanese agricultural treatises), vol. 4, Tōkyō, 1982, pp. 285–305;

For the first time, accurate illustrations of farming tools appear as 'vehicles of knowledge' in this treatise. The writings and drawings were designed to encourage the reader to copy these tools. However, as technical images, the drawings include surprisingly few details and give only few dimensions.

Therefore, the following questions arise:

1) How could one imitate and manufacture such tools without specific measurements?
2) What could be the reason for the lack of dimensions and measurements?

One explanation for the lack of dimensions and measurements could be that the author assumed that anyone interested in imitating and manufacturing the tools – probably a farmer – had the necessary knowledge to handle and process the respective materials. The author could obviously count on so-called tacit knowledge.

At this point, the special character of these agricultural treatises as a kind of textbook becomes visible; they address those already experienced in the field. For these people, one need not describe the nature of wood or iron or give measurements for the length of a handle or of a hoe, or for the angle between handle and blade.

The authors of such treatises wanted to show examples of (often new) devices so that a farmer, a rural carpenter, or a blacksmith could manufacture such tools based on his existing knowledge.

For example, although a number of dimensions are given for the plough, details such as the composition of the stand and drawbar or the attachment of the hook blade are not clearly visible. This device cannot be manufactured with these few data alone, so the experience of a professional carpenter was necessary.

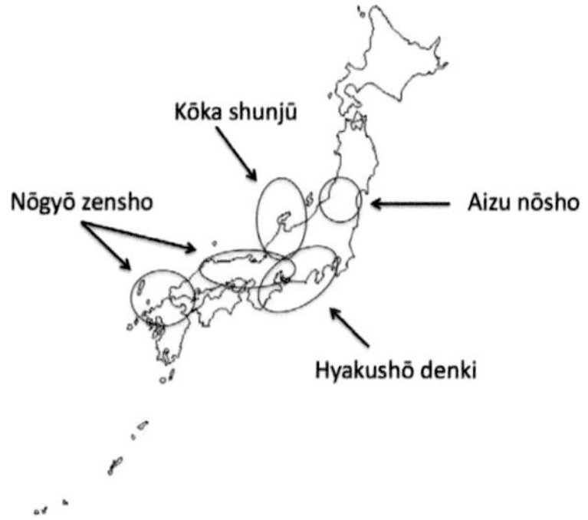

Fig. 4: Target audience of the four important agricultural treatises.

These treatises are all based on the author's local experience and do not claim to propagate the devices they describe for use in the whole of Japan. However, they remained essential for certain areas and promote agriculture there to a large extent.

From this map (Fig. 4) it is clear that even these few examples of agricultural treatises cover a large part of Japan. This also means that the dissemination of technical knowledge on a larger scale began as early as 1700.

These regional treatises were often copied, reprinted, amended, and usually served as models for other authors. The treatises often offered to the reader various options and examples (e.g. of equipment), from which he could select a tool convenient for him, or which appeared to be most useful for certain circumstances. Excerpts from the books or simplified versions became also widespread, contributing to a 'market for ideas' in this field.

Two kinds of technological change can be observed in seventeenth-century Japan. Looking at the various agricultural treatises, we see a technological change that was *experience-based*. We also see, especially in Miyazaki Yasusada's *Nōgyō zensho*, another kind of change. Miyazaki carried out what today we would call scientific experiments, e.g. applying systematic research into the use of materials, such as fertilizer. In doing so, he gained an understanding, based on practice, of why certain production techniques actually worked. The results of his research, which were published in his *Nōgyō zensho*, can therefore be called a *knowledge-based* technological change. With such an approach, he initiated a change that emphasized 'science' and 'technology' (to use modern terms), although others did not follow this path until many years later.

A step toward a more sophisticated dissemination of technical knowledge through drawings was made 120 years later by the agronomist Ōkura Nagatsune.[7] Having already published some general works on agriculture, Ōkura published the *Nōgu benri ron* (On the convenient use of agricultural tools) in 1822.[8] In contrast to the earlier treatises of this kind, which only presented devices from a single region, the author described and illustrated a plethora of devices from across the country, thus establishing what can

7 On Ōkura Nagatsune see Thomas C. Smith, 'Ōkura Nagatsune and the Technologists', in A.M. Craig and D.H. Shively, *Personality in Japanese History*, Berkeley: University of California Press, 1970. pp. 127–154.

8 For details on the *Nōgu benri ron* see the reprint in Nōsan gyoson bunka kyōkai (ed.), *Nihon nōsho zenshū* (A collection of Japanese agricultural treatises), vol. 15, Tōkyō, 1979.

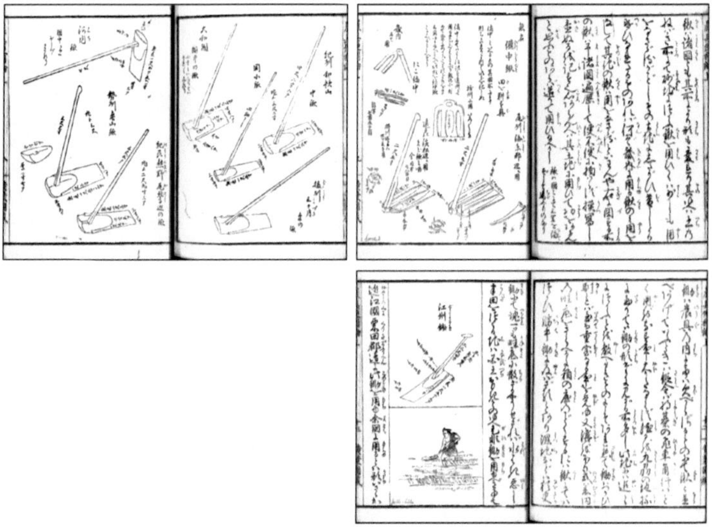

Fig. 5: Drawings of agricultural tools: various types of hoes (above) and the use of a spade (below) with dimensions and further details in the *Nōgu benri-ron* (1822). Source: Copy in the author's collection. See also Kikuchi Toshiyoshi (ed.), *Zusetsu Edo-jidai no gijutsu* (Edo-period technology, illustrated), Tōkyō, 1988, pp. 64–71.

once more be described as a 'market for ideas' with ideas given in the examples of tools (see Fig. 5).

In the illustrations (Fig. 5), the dimensions of the individual components are specified, as are the different types of materials used. Precise measurements, such as those for angles, and other technical details are also given. The author refers to the region in which each of these devices are used. He also adds comments on the type of soil for which they are most suitable and other details. This work actually promotes the dissemination and reproduction of knowledge on tools and devices tailored to serve practical purposes in various regions. The accuracy and details of the drawings vary from one section of the treatise to another.

Ōkura's *Nōgu benri ron* had a huge impact on Japanese agriculture. Because of its character as a textbook, a number of devices initially used only in certain areas became more widespread and were later employed nationwide.

Most of the drawings of farming tools still conform to traditional styles and look like 'pictorial illustrations'. However, in a gradual process the style changes and elements like dimensions etc.

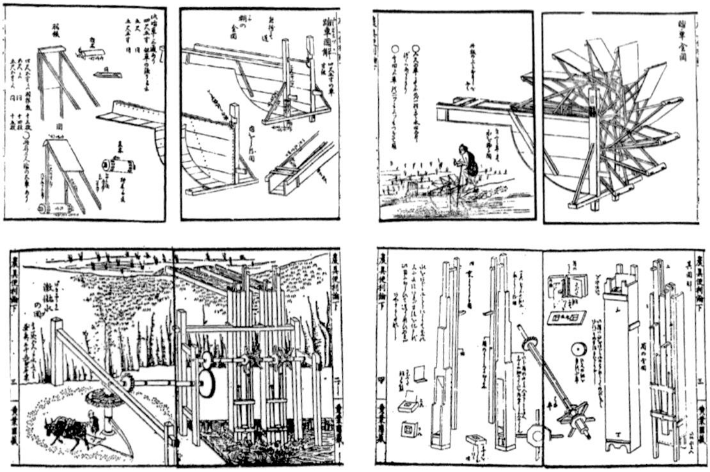

Fig. 6: Drawings of the irrigation machinery (water wheel and pomp) from the *Nōgu benri-ron* (1822). Source: Saegusa Hiroto (ed.), *Nihon kagaku koten zensho*, Tōkyō: Asahi shin-bun-sha 1944, vol. 11, pp. 35–202, pp. 140–143 and 160–164; see also Kikuchi Toshi-yoshi (ed.), *Zusetsu Edo-jidai no gijutsu* (Edo-period technology, illustrated), Tōkyō, 1988, pp. 156 and 170. Also Nōsan gyoson bunka kyōkai (eds), *Nihon nōsho zenshō* (A collection of Japanese agricultural treatises), vol. 15, Tōkyō, 1982, pp. 285–305;

are added, so that the results resemble more and more technical drawings.

But other parts of this treatise give explanations of more complex devices such as water wheels and pumps, and thereby come close to a real 'technical text book' in the Western sense (Fig. 6).

In addition to an overall view of the device, which gives an impression of its use, the illustrations provide details that include dimensions of individual components, and the kind of wood suitable for each part of the device. With such specific information, a skilled farmer or rural craftsmen could easily build such a device. Agricultural treatises are a good example of how technical drawing methods developed over time and thus nicely illustrate the development of 'how-to' books, how such books facilitated the spread of knowledge, as well as how technical drawing methods advanced over time.

4. BOOKS AND DRAWINGS IN CRAFTS

Interestingly, the situation in the craft sector is quite different from that in agriculture. There are some rare drawings from the late sixteenth century represent certain craft techniques pictorially, but which do not serve as carriers of knowledge about these techniques or about how to process the materials involved. The only exceptions are architectural drawings. The building trade traditionally worked with floor plans and movable parts, which developed over time from two-dimensional layouts to plans with turning flaps that provided the viewer with a vivid three-dimensional depiction of the resulting building. From the late fifteenth century, such three-dimensional plans appeared more and more, especially those used for official or manorial buildings, shrines, temples, and so on. Unlike blacksmiths, hoe-manufacturing carpenters, sieve makers, and basket makers, who mostly worked alone, several people usually had to work together in the building trade. Plans therefore had to be accessible and comprehensible to various persons. Drawings illustrating certain construction techniques became increasingly common.

Another tool for passing on technical details in the building trade are so-called *Hinagatabon* (lit. Books on models), a book genre containing architectural drawings of models, templates of shrines or temple buildings, and sample plans (see Fig. 7a). These templates, as well as the floor plans with movable parts mentioned before, do not constitute a 'textbook' in the narrow sense, although they could be 'read' and put into practice by a trained craftsman.

In addition, there are also detailed manuals that explain the construction of houses, walls, roof trusses, and so on. They are comparatively complex and widespread (Fig. 7b).

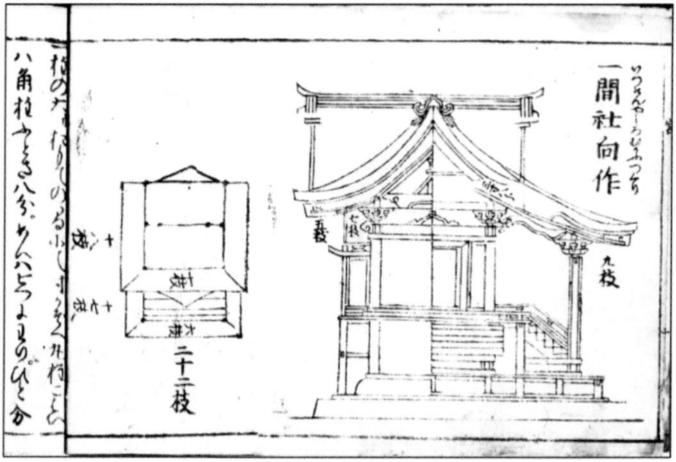

Fig. 7a

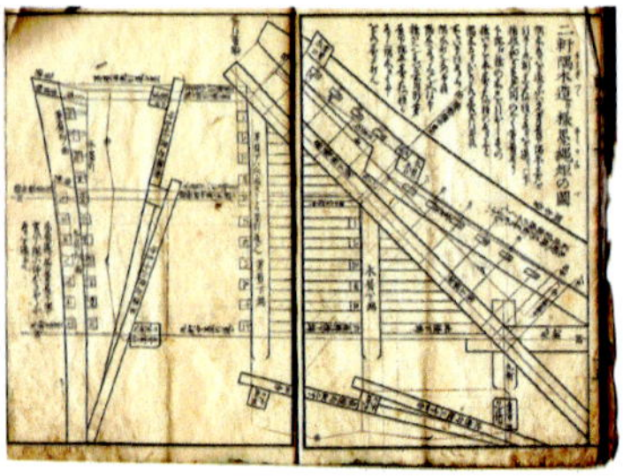

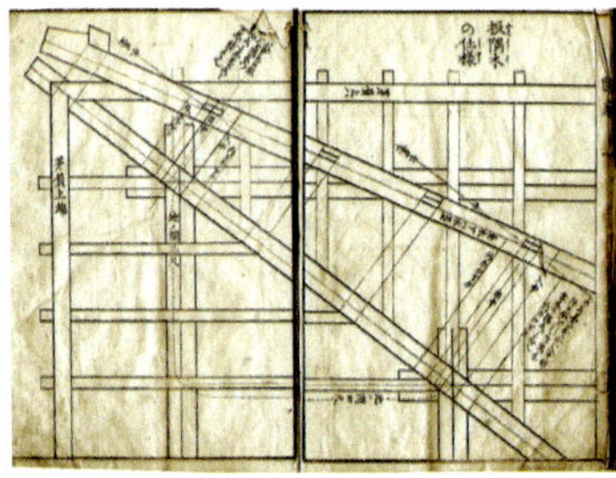

Fig. 7a & Fig. 7b

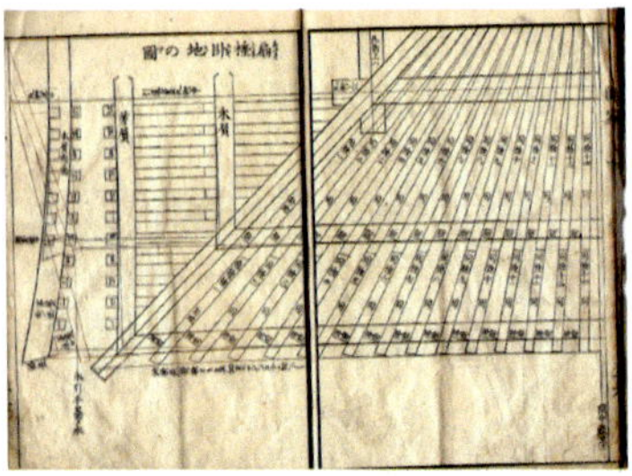

Fig. 7c

Fig. 7a–7c: Above: A template for a shrine building with textual information from a *Hinagata*-booklet titled *Shinpan Miya hinagata 1* (Models of shrines, New Edition Part I) dated 1825; below: Templates showing details of a roof construction from a carpenter's manual titled *Shoka hinagata*, (1853). Source: Copies in the author's collection.

As the building trade required the circulation of written (technical) documents, it was difficult to limit the transfer of knowledge. However, for most crafts, practically no textbook or instruction manual, which could disseminate detailed technical knowledge either to apprentices or to the public, existed from the seventeenth to the nineteenth century.

In the early Edo period, crafts and their techniques were typically considered to be the responsibility of the artisan's family and were often regarded as a secret not to be passed on to others. This situation also reflects the Japanese way of learning in pre-modern times, which was akin to an apprenticeship, whereby the instructor would give a one-to-one demonstration before giving one-to-one feedback. This was 'learning by doing' and 'learning through experience'. Technical knowledge, mostly practical knowledge based on experience, was passed down orally through the generations, while pictorial representations played only a marginal role and rarely survived.

An extremely rare example of a manual, perhaps better called a 'collection of sketches', has survived in the family of a former carpenter in mid-Japan near the lake Suwa in Nagano prefecture.

This *Korai sōden* (lit. Ancient inheritance),[9] which was probably written around 1789, gives us a glimpse into how a carpenter made detailed notes and sketches, not only for himself, but also for his apprentices or successors. The small 60-page booklet was passed on as a *hidensho,* a carpenter's script with secret traditions. It contains various sketches of tools, devices, and signboards, and describes other objects made of wood. It gives dimensions and additional references. The different objects are simply collected and described one after the other, and no order or sequence is identifiable.

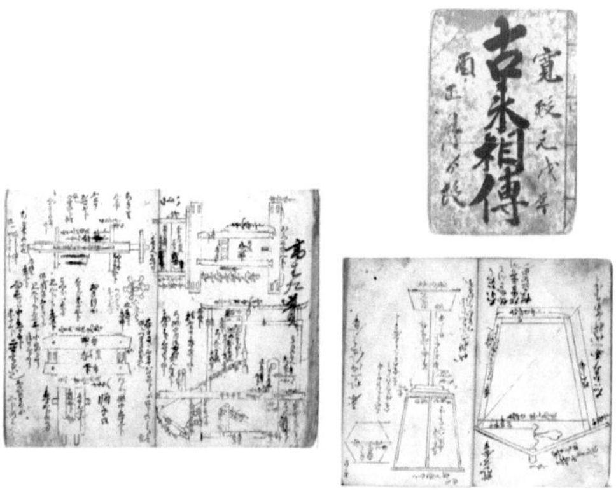

Fig. 8a

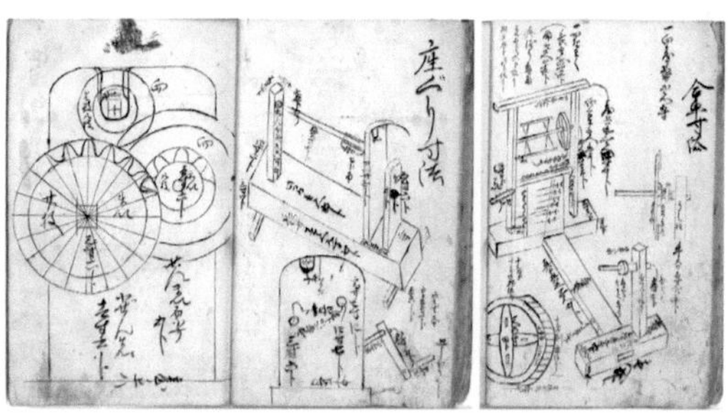

Fig.8a and 8b: Cover and examples of sketches of various tools and devices in the *Korai sōden* (Ancient Inheritance), (Templates with permission by Fujimori Yuichi). Most of the roughly depicted tools and machines are used for silk processing.

9 For *Korai sōden* (Ancient inheritance) see Okuda Masamitsu, 'Korai sōden ni tsuite (shōkai)' in Tabako to Shio no Hakubutsukan (ed.), *Edo no mekanizumu*, Tōkyō 1989, pp. 115–117.

One drawing (Fig. 8a) describes on the left different objects, which are not necessarily related to each other. The other (Fig. 8b) shows a device used in a farmer's household to prepare yarn. An outside crank is shown with a description, as well as two cogwheels. In the middle there is a sketch of a spinning device called *zaguri*. It is a typical tool in cloth making, which virtually every small producer needed.

All of these sketches are quite rough and provide only a few details and brief explanations, and were evidently designed for someone already possessing significant related knowledge. For example, a rural carpenter with some experience in tool making should be able to make such a device according to this given sketch. All the drawings were made by hand using a brush, and were not printed. Different handwritings are distinguishable, suggesting that subsequent generations may have made amendments to existing sketches and passed them on to their successors.

Secret sketchbooks or notes of this kind were used for various crafts where the measurements and details necessary for making objects could only be transmitted by a sketch on paper. Another example (Fig. 9) is from a book by the famous gunsmith Kunitomo Ikkansai from 1818. Certain information, such as the method for manufacturing the rifle barrel, was not intended for publication and was used only in the workshop.

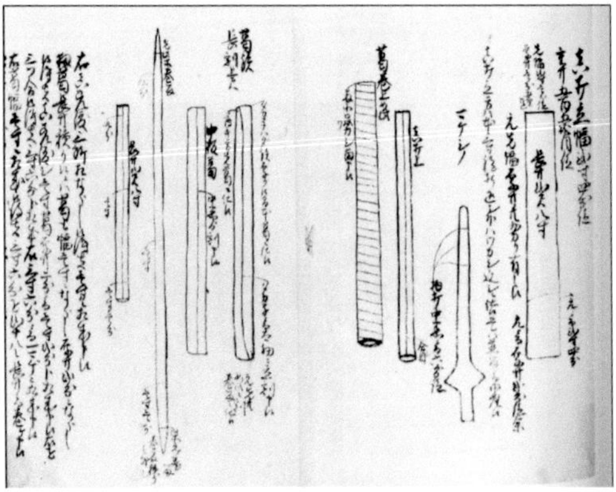

Fig. 9: Sketches for the production of a rifle barrel by Kunitomo Ikkansai (1818). Source: Shiritsu Nagahama rekishi hakubutsu-kan (comp.), *Kunitomo teppō kaji – sono sekai –* (The world of the Kunitomo gunsmith), Nagahama Rekishi hakubutsu-kan, Catalogue of an Exhibition, 1985, p. 45.

There are, however, examples whereby technical knowledge was disseminated to people who were not necessarily involved in manufacturing technical devices, but who only used them. Technical knowledge thereby began to spread beyond the boundaries of craftsmanship to the more general public, as the following example shows.

The *Kaisen anjoroku* (On safe sea-travel) dated 1810 and written by Hattori Yoshitaka, is a record of a steerman responsible for shipping a yearly tribute from Hachijōjima (part of the Izu archipelago) to Edo, and for transporting various goods back to the island. The journey from the island, 200 km south of the Izu peninsula and nearly 300 km from Edo, was not easy and sometimes quite dangerous. It was for this reason the author wrote this report for his successors. It gave a record, not only of the routes to travel, but also provided various details about the ship, such as terms concerning the ship's hull (Fig. 10), details of the ship's rudder, ropes, anchor, compass, etc.

This report was not designed for a ship carpenter, but for someone who was probably a skipper who had interest, not only in sailing the ship, but also to understand more about the vessel for which he was responsible.

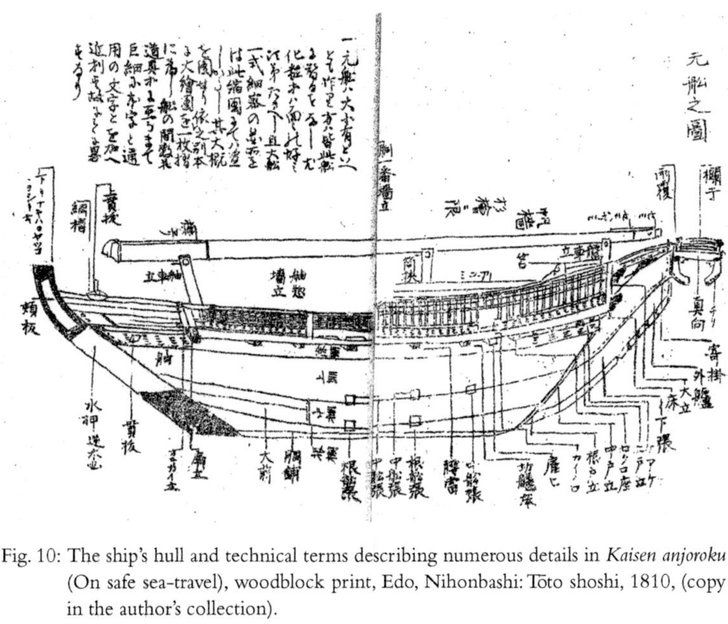

Fig. 10: The ship's hull and technical terms describing numerous details in *Kaisen anjoroku* (On safe sea-travel), woodblock print, Edo, Nihonbashi: Tōto shoshi, 1810, (copy in the author's collection).

This is just one example of such books, which contained technical templates but were not part of the secret tradition within a craft. Aiming at a broader public, they mark a turning point in the spread of technical knowledge during the last years of the eighteenth and early years of the nineteenth century. Instead of keeping technical knowledge secret, e.g. restricting this kind of sensitive military technical knowledge to the members of the Kunitomo family, the publication of technical details and easily available information about this kind of technical knowledge became increasingly commonplace.

Only a few years earlier another astonishing book appeared on the Japanese market, seemingly without any precedent. *Karakuri zui* (Illustrated machinery), a technical textbook that is exemplary of its kind, is a three-volume work by Hosokawa Hanzō Yorinao published in 1796.[10] Comprising a single volume, it contains sufficiently precise instructions and drawings for the construction of different automata and, importantly, it is also explicitly a manual for their reproduction.

Automata of the kind shown in this book were already very popular during the seventeenth and eighteenth centuries, but no records existed showing their construction or advising how to construct them, as this knowledge was probably kept secret.

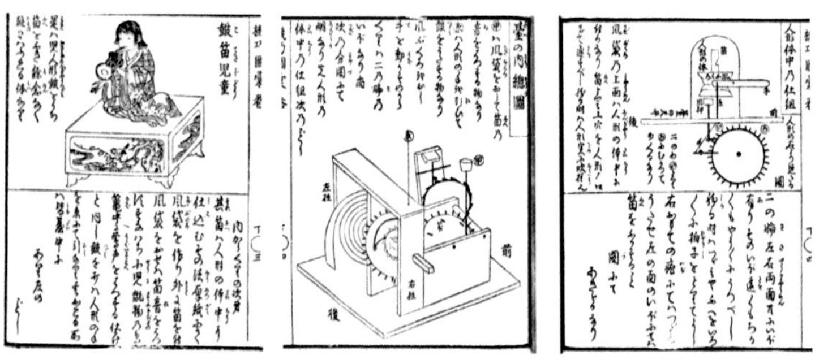

Fig. 11: Toy acrobat. Driven by a spring mechanism, the figure's hand hits a drum, while inside, flute tones are produced with the help of a simultaneously driven bellow. Source: Kōchi-ken Nankoku-shi-ritsu kyōiku kenkyū-sho (eds), *Karakuri-zui* (Illustrated machinery) vol. 3, Nankoku, Kōchi pref., 1995 (annotated reprint), pp. 79–81.

[10] For details on Japanese automata see Tatsukawa Shōji, *Karakuri*, Tōkyō: Hōsei daigaku shuppan-kai, 1969 (various editions following) and especially on the *Karakuri zui* on pp. 131–178; also Fukumoto Kazuo, *Karakuri gijutsu-shi-wa* (Essays on the *karakuri*-technique), Tōkyō: Fuji shuppan-sha, 1982; also Erich Pauer, 'Japanische Automaten' (Japanese automata), in *Technikgeschichte*, vol. 77 (2010), 4, pp. 321–351.

No written accounts are known of prior to 1796, when *Karakuri zui* was published.

In a first volume, quite simple automata are explained, e.g. an acrobat and a puppet boy who beats a drum (Fig. 11, left). The movement of the hand is powered by a spring mechanism and is clearly visible (Fig. 11, middle). A cross-section through the device is given on the right (Fig. 11, right).

It should be emphasized that *Karakuri zui* is not just an instruction manual for toys. Fundamental principles of mechanics, such as the principle of springs and other mechanical details, are explained explicitly. This becomes evident in the next example (Fig. 12 top) in which an acrobat performs a series of slow-

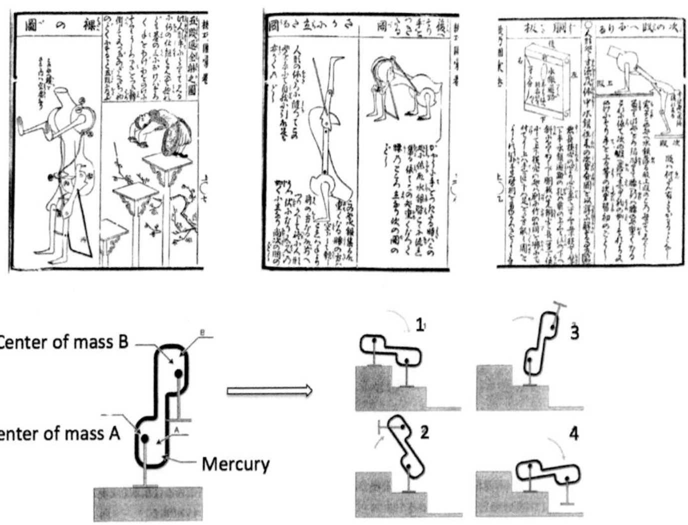

Fig. 12: 'Acrobat'-automaton doing slow-motion back-somersaults (above). Source: Kōchi-ken Nankoku-shi-ritsu kyōiku kenkyū-sho (eds), *Karakuri-zui* (Illustrated machinery), Nankoku, Kōchi pref., 1995 (annotated reprint), pp. 51, 53 and 55.

The function of the doll is clear from the diagrams (below): Inside the doll there is a hollow body standing vertically in the resting state with two chambers (A, B) at both ends and a narrow connecting piece (bottom left). In the lower chamber A there is mercury. Briefly pushed, the doll falls over the back (1) and then rests on the next step (2), while the mercury slowly (!) runs from chamber A, which is now at the top, into chamber B. The mercury is then released into chamber B by the next step (2). The process is repeated over the further steps, since the swing from the first rollover now causes the feet to swing up and forward (3) and the shift of the centre of gravity causes another rollover to the next step (4). (How it works is shown using a modified illustration from Minesaki Sōgo in *Robokon magajin* (Robocon Magazine) No. 3 (1999), p. 113).

motion back-somersaults down a staircase. Previously, it had been a secret how the mechanism worked, and this riddle could not be solved by the audience at that time. *Karakuri zui* revealed the mechanics, and therefore the 'secret' for the very first time to a broader audience, which was truly revolutionary at the time.

The figures (Fig. 12 bottom) show what is happening. The doll-automaton has two containers, A and B, which are connected by a thin tube. The lower part A contains mercury. A short push of the upper body of the doll makes the mercury start moving through the narrow tube into container B, thereby causing a shift in the centre of gravity and the doll to roll over on to the step below. Momentum causes the doll to continue down the staircase.

Both mechanical and physical principles, such as the shifting centre of gravity, are thereby explained. That, among other factors, is what makes *Karakuri zui* an outstanding book and landmark in the history of technology in Japan.

If these machines are already comparatively complicated in construction, the author Hosokawa goes one step further. The most famous example of a more sophisticated automaton is the 'Tea-serving

Fig. 13: The 'Tea-serving boy' (an automaton) on stage (eighteenth century).
Source: Kikuchi Toshiyoshi (ed.), *Zufu – Edo jidai no gijutsu – jō* (Illustrated history of Edo period, vol. 1), Tōkyō: Kōwa Shuppan 1988, p. 616.
The use of a 'Tea-serving boy' (center) in courtly society from the *Ehon kiku kasane* from the mid-eighteenth century. Two simple automatons are visible in front of the figure sitting on the left, behind the person kneeling on the right a third automaton in a container.

boy'. The puppet – roughly 50 cm in height – runs on stage (Fig. 13) in the midst of a group of nobles. It holds a bowl of tea on a tray with both hands presenting it to one of the guests, who takes the cup from the tray, drinks the tea, then puts the cup back in the hands of the puppet, which then turns around and returns to the host. The mechanism by which this is accomplished is not visible to the audience.

Karakuri zui contains detailed technical drawings of every part of this automaton with certain details on materials and material processing, and with specifications concerning material (different kinds of wood for different parts) and construction details. Looking at the various detailed figures in the book (Fig. 14), one gets the impression, that an interested and sufficiently skilled craftsman should have been able to manufacture such an automaton. But contrary to the aim of serving as a model for reproduction explicitly stated in the book, this was not the case.

At least one drawing is not correct, and it is not clear whether this was done on purpose by the author or whether it was simply a mishap by the carver creating the woodblocks. With the help of the drawings in the book alone, it was not possible to construct this automaton.[11] However, this does not seem to have been too much of a problem for contemporaries, because the craftsmen were able to inquire through personal contacts or study existing

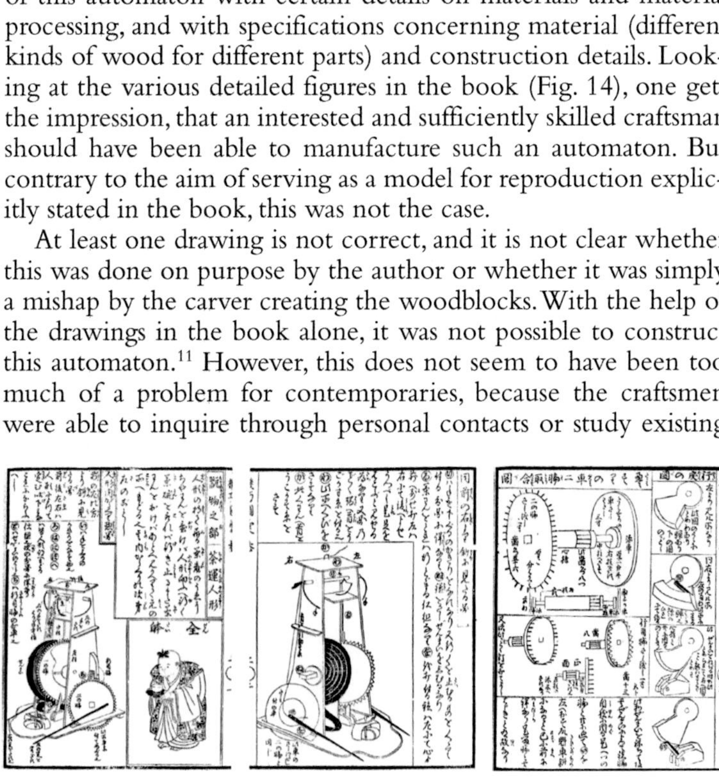

Fig. 14: Three (of a total of 12) pages of the building instructions for the "Tea-serving Boy," an automaton described in the *Karakuri zui* from 1796.
Source: Kōchi-ken Nankoku-shi-ritsu-kyōiku kenkyū-sho (eds), *Karakuri zui* (Illustrated machinery), Nankoku, Kōchi pref., 1995 (annotated reprint, pp. 39, 40, 41 and 49.
The illustration is to be 'read' from left to right. The construction manual begins on the right with an overview of the mechanics of the figure and continues with the construction and the individual parts.

11 This mistake was only discovered in the 1970s; for details see Pauer, 'Japanische Automaten', pp. 335–336.

examples of the automaton (some of them still preserved in museums today), and thus close gaps in their knowledge.

There exists further evidence that these technical details did not remain a secret. The sketches shown in Fig. 15 produced by Ono Benkichi (1801–1870), a famous inventor and technician from Kanazawa, were based on the drawings in *Karakuri zui* and on experience with a working model. Ono Benkichi used these sketches to construct such automata as the 'Tea-serving boy' and other *karakuri*-devices in his local area.[12] He is also famous for his work as photographer.[13]

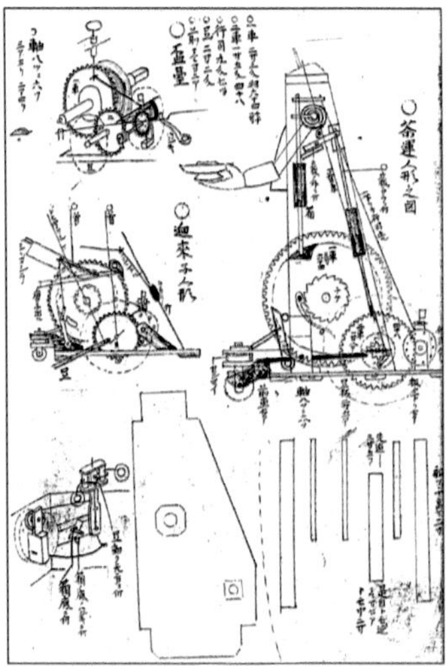

Fig.15: Construction details of the "Tea-serving boy" by Ono Benkichi based on the *Karakuri zui* (Illustrated machinery) and a working model (sketches made before 1850). Source: Kanazawa Karakuri Memorial Museum, Kanazawa.

[12] Detailed descriptions of the various attempts to manufacture this 'Tea-serving boy' are described in Pauer, 'Japanische Automaten', pp. 336–337.

[13] Around ten 'Tea-serving boy' automata, mostly made in the first half of the nineteenth century, have survived in various Japanese museums. One 'boy', probably made by Hosokawa Hanzō himself, is still to be seen in the Kōchi-kenritsu rekishi-minzoku shiryō-kan (Kōchi prefectural Museum for History and Ethnology), Nankoku, Kōchi prefecture; another one is preserved in the collection of the Karakuri kinen-kan (Karakuri Memorial) in Ono near Kanazawa, Ishikawa prefecture.

The basic technique for making such automata derived from the construction of clocks, but we have no concrete details of the historic process involved. Clocks already arrived in Japan in the late sixteenth century via China. Japanese craftsmen were first taught by Jesuits in Nagasaki, who had also taught Chinese craftsmen in Beijing. In this way clock-making techniques then spread all over Japan. However, no charts, manuals, or books on clock making are known before *Karakuri zui* was published.

One volume of *Karakuri zui* is dedicated exclusively to the reconstruction of a clock. The drawings from *Karakuri zui* (Fig. 16) do indeed appear to be technical and engineering drawings. In addition to various sectional drawings (top view, front view, and side view), we find perspective drawings. Moreover, numerous individual drawings explain the functions of certain components.

Japanese historians of technology have repeatedly called *Karakuri zui* the very first technical book in Japan. It was distributed widely, partly copied, re-written, and amended by other authors and therefore did much to disseminate technical knowledge. Looking at the technical drawings, the construction charts of the

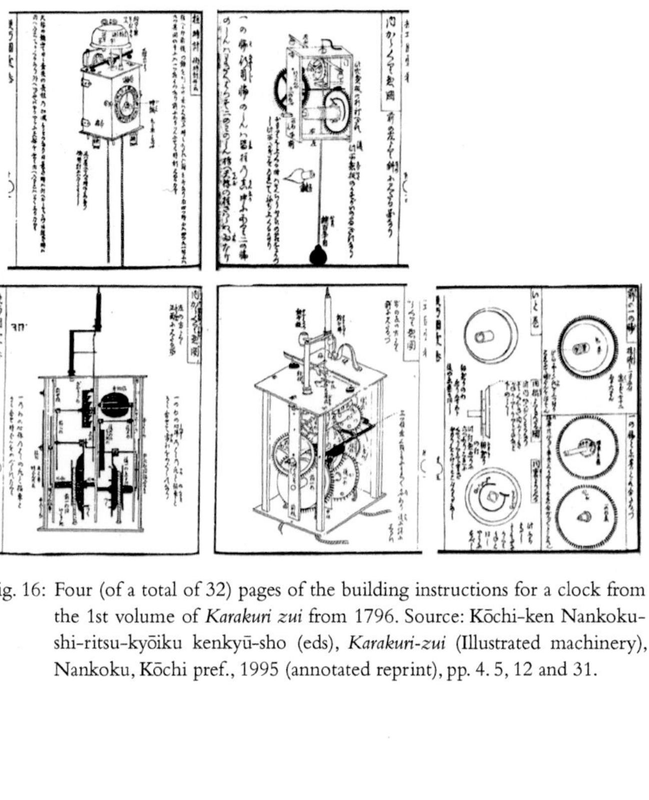

Fig. 16: Four (of a total of 32) pages of the building instructions for a clock from the 1st volume of *Karakuri zui* from 1796. Source: Kōchi-ken Nankoku-shi-ritsu-kyōiku kenkyū-sho (eds), *Karakuri-zui* (Illustrated machinery), Nankoku, Kōchi pref., 1995 (annotated reprint), pp. 4. 5, 12 and 31.

clock and the construction details of the 'Tea-serving boy', these are already very close to Western practices. Therefore, one can say that with this *Karakuri zui* Japanese technicians had reached a level of drawing that enabled highly skilled craftsmen to also understand European drawings that came into the country in the middle of the nineteenth century in great numbers.

5. CONCLUSION

'How-to books' developed in Japan mostly in agriculture from the seventeenth century onward. Many of these books contain texts and pictorial illustrations. Over time, visual representations became formalized and codified and changed from pictorial illustrations to more abstract drawings of technical tools and devices. In this way, people in Japan acquired, not only technical knowledge for their work, but also knowledge of how such representations should be 'read' and how the relevant tools and devices should be constructed. At the end of this process, eventually the technical drawings were often more important than the texts and, in some circumstances, became the sole source of knowledge transfer.

One question remains: what was the reason for the change from books with illustrations that were often not connected to the text, as was usual in earlier times, to books in which text and illustrations corresponded and the technical drawings gave more insight into the accompanying text and vice versa?

One point should be emphasized: more sophisticated technical drawings appeared in various fields, not only in agriculture or in clock making, around the year 1800. A possible explanation for such a change may be found in the fact that, following the lifting of the ban on the import of Western books in 1720, a great number of Western books poured into Japan in the second half of the eighteenth century and were studied in several places in different domains. These books, especially those on medicine, which gave detailed insights into the human body, might have been a vehicle for knowledge about how to draw precise illustrations from various angles (or, in the case of medicine, of parts of the human body) to accompany an explanatory text. Authors and readers thereby gained a deeper understanding of the body – or, in other cases in a more technical sense, of a device and more complex machinery. It is possible that imitation of these drawings led to the appearance of more Western-style technical drawings. Examples of drawings used for clock making, constructing waterwheels and pumps etc., to scale, with dimensions, weights, materials, notes on the processing of the material, application of the final device and so on, illustrate this development, which appeared around 1800, must therefore be called a turning point in the drawing of technical devices and machinery.

If one now takes a look at the development of drawing over the centuries (except for the drawings from the eighth century, which obviously had no traceable impact), it becomes clear that early drawings can initially be characterized as pure 'depiction'. Even with the simplest form of representation, it was clear to the reader which device was represented. However, increasing mechanization and complexity of individual devices meant that drawings did not only 'represent' a device pictorially, but were also templates to 'construct' or to 'manufacture' the device shown. The more complicated constructions and devices could no longer be manufactured by ordinary people, such as farmers, and needed expert knowledge; as specialists became more common there were new requirements for the ability to 'read' drawings. This led to the development of a style of representation that, perhaps due to the import of foreign works from various disciplines, approached the Western style.

There could also be an additional influence; that of Western painting. In the last third of the eighteenth century, geometric perspective was already being studied in Japan,[14] the influence of which was clearly reflected in the depictions of the clock and automatons of Hosokawa Hanzō in 1796.

In the first decades of the nineteenth century, when a larger number of Western technical manuals, which provided guidelines for the construction of blast furnaces, reverberatory furnaces, canals, fortifications, textile machinery, and even ships and steam engines, were imported to Japan, knowledge of how to 'read' such technical drawings became one of the pillars for understanding Western technology. To take just one example, in the 1860s a Japanese engineer named Tanaka Hisashige (1799~1881) was able to build and run a small-scale steam locomotive solely based on imported technical drawings without ever having seen a steam locomotive running in reality. Technological development, as seen in the way that technical illustrations have changed over 250 years, has paved the way for the technological modernization of Japan.

---

[14] See Kobayashi-Sato Yoriko, 'An assimilation between two different cultures: Japan and the West during the Edo period', in: Michael North (ed.), *Artistic and Cultural Exchanges between Europe and Asia, 1400–1900*, Farnham: Ashgate 2010, pp. 165–186, esp. p.166.

REFERENCES:

Fukumoto Kazuo, *Karakuri gijutsu-shi-wa* (Essays on the *karakuri*-technique), Tōkyō: Fuji shuppan-sha, 1982.

Furushima Toshio, *Furushima Toshio chosaku-shū* (Collected works of Furushima Toshio), Tōkyō: Tōkyō daigaku shuppan-kai, 1975.

Iinuma Jirō, *Nihon no nōsho* (Japanese agricultural treatises) (Chūkō shinsho 852), Tōkyō: Chūō-kōron-sha, 1987.

Iinuma Jirō, *Kinsei nōsho ni manabu* (Learning from the agricultural treatises of the early modern period) (NHK bukkusu 271), Tōkyō: Nihon hōsō shuppan kyōkai, 1976.

Kaisen anjoroku (On safe sea-travel), woodblock print, Edo, Nihonbashi: Tōto shoshi, 1810, (copy in the author's collection).

Kobayashi-Sato Yoriko, 'An Assimilation between two Different Cultures: Japan and the West during the Edo Period', in Michael North (ed.), *Artistic and Cultural Exchanges between Europe and Asia, 1400–1900*, Farnham: Ashgate, 2010, pp. 165–186.

Nōsan gyoson bunka kyōkai (ed.), *Nihon nōsho zenshū* (A collection of Japanese agricultural treatises), Tōkyō: Nōsan gyoson bunka kyōkai, 1978.

Ōbayashi-gumi (ed.), *Kodai Izumo-taisha no fukugen* (The reconstruction of the ancient Izumo shrine), Tōkyō: Gakusei-sha, 2000.

Ōbayashi-gumi purojekuto chīmu (publ.), *Yomigaeru kodai dai kensetsu jidai – kyodai kenzōbutsu wo fukugen suru* (Reviving the ancient period of large-scale architecture – reconstructing giant buildings), Tōkyō: Ōbayashi-gumi, 2002.

Okuda Masamitsu, 'Korai sōden ni tsuite (shōkai)'(An introduction to the Korai sōden), in Tabako to Shio no Hakubutsukan (ed.), *Edo no mekanizumu*, Tōkyō: Tabako to Shio no Hakubutsukan, 1989.

Ono Takeo, *Aizu nōsho* (Agricultural treatise of Aizu), Tōkyō: Itō shoten, 1944.

Pauer, Erich, *Technik-Wirtschaft-Gesellschaft. Der Einfluss wirtschaftlicher und gesellschaftlicher Veraenderungen auf die Entwicklung der landwirtschaftlichen Geraete in der vorindustriellen Epoche Japans ab dem 17. Jahrhundert* (Changes in economy and society and the development of agricultural tools in pre-industrial Japan), Wien: Institut für Japanologie, 1973, (Beiträge zur Japanologie vol. 4).

Pauer, Erich, 'Japanische Automaten' (Japanese automata), in *Technikgeschichte*, vol. 77 (2010), pp. 321–351.

Robertson, Jennifer, 'Japanese Farm Manuals: a Literature of Discovery', in: *Peasant Studies*, vol. 11, no. 3 (Spring 1984), pp. 169–194.

Smith, Thomas C., 'Ōkura Nagatsune and the technologists', in A.M. Craig and D.H. Shively (eds), *Personality in Japanese history*, Berkeley: University of California Press, 1970.

3

Dissemination of Knowledge and Technology: The Extensive Range of Exhibitions in Japan in the Eighteenth and Nineteenth Centuries

ITŌ Mamiko

1. INTRODUCTION

THE FIRST DOMESTIC exposition organized by the Japanese government was held in Tōkyō in 1877. Despite the ongoing Satsuma Rebellion, a revolt of disaffected samurai against the new Meiji Government, the First National Industrial Exhibition was declared open by Ōkubo Toshimichi (1830–1878), the first Minister for Home Affairs, on 21 August. The launch marked the decision by the government to put greater emphasis on industry than had been the case in the first decade after the Meiji Restoration.

There were five national industrial exhibitions held during the Meiji era (1868–1912). One of the aims of holding these exhibitions was to stimulate Japanese submissions to international expositions. Exhibitors could not submit co-produced objects outside the prefecture, neither to the national exhibitions nor to international expositions, until the St. Louis Exposition in 1904. Besides the national industrial exhibitions, there were competitive exhibitions called *kyōshinkai* (meetings for collective advancement), which were held in the prefectures with the support of the government. These were more or less product trade fairs specializing in one category, and as such generated greater competition among craftsmen than the national exhibitions. These exhibitions gave rise to commercial museums, just as the experience of participation in international exhibitions gave rise to national museums.

The government also established schools for advanced science and technology and model factories in order to promote new industries. For example, the Ministry of Engineering founded Japan's first national technical school, the Kōgakuryō, in 1873 and Akabane Engineering Works in 1871.[1]

However, these education measures were aimed at a young and still largely inexperienced generation of students. The new engineering factories were not intended for indigenous craftsmen. Exhibitions, on the other hand, provided a practical and immediate possibility to acquire knowledge, to then create something innovative and submit it to a subsequent exhibition. Exhibitions were a useful medium for the dissemination of new techniques.

The exhibitions may have been envisaged as a medium of dissemination, but they also proved to be a huge entertainment draw. Visitors enjoyed the technological innovations and the exotic entries, having been quite familiar with the concept of exhibitions through *yakuhin-e* (*yakuhinkai*) (medical and pharmaceutical shows) or *bussan-kai* (trade shows), which had first been held in the middle of the eighteenth century, and which had flourished in the first half of the nineteenth century. The government stressed the difference between the new *hakurankai* exhibitions and the traditional *yakuhin-e* exhibitions.

This chapter recounts the history of exhibitions in the eighteenth and nineteenth centuries and describes how the government used exhibitions as a medium for the dissemination of new knowledge and technologies among visitors of a wide range of social and age groups. First, it outlines the origins of the exhibitions and conferences (*yakuhin-e/yakuhinkai*) in the development of natural history in the Edo period, and then investigates the emergence of *hakurankai* exhibitions, as well as domestic exhibitions and national museums, which resulted from Japan's decision to participate in the international exposition in Vienna in 1873. Finally, it illustrates the role that amateur scholars played in this process. In light of the evidence, it appears that inquisitiveness in itself lead to a new style of research beyond rank and class, which sometimes connected professionals and non-professionals. This

---

1 Miyoshi Nobuhiro, *Nihon kōgyō kyōiku seiritsu-shi no kenkyū* (Studies on the formation of technical education in Japan), Tōkyō: Kazama shobō, 1979; Uemura Shōji, 'Meiji shoki kōgaku kyōiku kikan no setsuritsu: Kōgaku-ryō ni tsuite' (The establishment of the Imperial College of Engineering in the early Meiji period), in *Shakai kagaku*, vol. 40, no. 3, 2010, pp. 21–47.

chapter describes the important role that amateur scholars played in laying the foundation for the introduction of new or foreign academic disciplines.

2. EMERGENCE OF EXHIBITIONS IN JAPAN

In Japan, natural history developed between the seventeenth and the nineteenth centuries. This development during the Edo period (1603–1868) influenced not only scholarship, but also culture at large. In the hundred years from the middle of the eighteenth century, natural history had wide appeal in both Europe and Asia.[2] In stark contrast to European countries, however, natural history in Japan did not develop through overseas expeditions. Under the foreign policy of the shogunate, Japanese individuals were not permitted to travel overseas during this era. Even imported information was suspect; only Chinese and Dutch merchants were allowed into Japanese ports between 1639 and 1854, and their influence and movement were monitored and curtailed. Only Satsuma domain in southern Kyūshū was allowed to trade with the Ryūkyū Islands (Okinawa). This was a source of Chinese goods throughout the Tokugawa era.

The Tokugawa shogunate maintained important economic and political relations with Korea via Tsushima domain, which derives its name from the small island located about halfway between Kyūshū and southern Korea. Although the policy prevented exploration abroad, the Japanese still enjoyed a flourishing culture, which included an interest in natural history, and this interest spread not only among intellectuals and wealthy aristocrats, but also among the general public

Natural history in Japan prospered in the eighteenth century. The political background of this popularity were the Kyōhō Reforms in the years between 1716 and 1736 introduced by

2 The following draws on Lynn Barber, *The Heyday of Natural History, 1820–1870*, London: Jonathan Cape, 1980, and Lynn L. Merrill, *The Romance of Victorian Natural History*, Oxford University Press, 1989. On Japan, Nishimura Saburō, *Bunmei no naka no hakubutsu-gaku: Sei-Ōto Nihon* (Natural history in civilizations: The West and Japan), Tōkyō: Kinokuniya, 1999, gives a useful introduction. Imahashi Riko, *Edo no kachō-ga hakubutsu-gaku wo meguru bunka to sono hyōshō* (Birds and flowers: The representation of natural history during the Edo period), Tōkyō: Sukaidoa, 1996, focuses on natural history in the Edo period from the perspective of literature and painting.

the eighth shogun, Tokugawa Yoshimune, mainly to promote financial reconstruction. One of the new policies of the government was the development of a domestic pharmacopeia, due to the fact that the importation of foreign medicines was putting a strain on silver and copper reserves. Throughout Japan, the search began for new species or subspecies, and the government promoted the commercialization and development of drugs. In 1721, Tokugawa Yoshimune commissioned the nationwide exploration and collection of previously unknown medicinal materials; in 1734 he launched another comprehensive project ordering investigation not just into pharmacopeia, but also into any natural production that promoted local industry. This was part of Yoshimune's plan to stimulate new industries and increase national wealth. At that time, there was a threat of famine in Japan, and the investigation was one of the measures undertaken to counter this threat and to encourage new industries.

Tokugawa Yoshimune's new policies also led to a transformation of Japanese natural history. New species or subspecies were sought as substitutes for foreign ingredients in medicinal recipes; scholars often met and discussed identification and nomenclature, and elaborated on their debates in books. Over time, they organized various societies and study groups and held academic conferences on medicine.

The scholars held conferences to discuss materials not found in textbooks. As an example, in 1757 Tamura Ransui (1718–1776), who was a doctor, botanist and herbalist at the shogunal medical academy in Edo (*Edo-Igaku-kan*), organized the first *'yakuhin-e'* conference. There he exhibited rare materials that he owned in order to compare, observe and discuss them with his pupils. One of his pupils, Hiraga Gennai (1728–1780), a scholar of natural history and a polymath, then expanded the conference to become an annual exhibition. The fifth and last of these, held in 1762, was the biggest, and included not only pharmacopoeia, but also natural products. In addition to new discoveries and natural products from distant lands, Hiraga Gennai also wanted to gather domestic items. Scholars from all over Japan attended this showcasing of rare items, and those who could not travel to Edo for the exhibition were able to obtain items through the network of pharmacies. Hiraga Gennai had distributed nationwide advertisements for submissions six months before the event, asking exhibitors for natural rarities that had not been previously displayed.[3] The exhibitors, which included drug sellers, herbalists, and amateur scholars (collectors), brought their exhibits to the nearest of twenty-five agencies across the country. The agencies then sent these exhibits to five famous wholesale drug companies and gardeners in the urban centres of Edo, Kyōto, and Ōsaka, who in turn sent them to Hiraga Gennai. The objects would be returned to the exhibitors by the same channels. The organizers also financed the round-trip transportation costs. The network of herbalists, botanists, scholars, pharmacies, and non-professionals across the country had already reached a level of sophistication that would make this complex movement of objects possible.

Yakuhin-e exhibitions opened up new research methods to scholars. With the increasing interest in objects, people began to collect real objects and exchange knowledge through exhibits and conferences (*yakuhin-e/yakuhinkai*). There were rare objects or animals (sometimes taxidermied) to facilitate research of unfamiliar subjects. Such exhibits were open to the public free of charge and, consequently, drew substantial crowds. Contemporary records show that the exhibitions were extremely popular (Fig. 1).[4]

Fig. 1: The *yakuhin-e* exhibition at the medical school in the province of Owari. Source: Igaku-kan yakuhin-e in *Owari meisho zue, zenhen* 2 (Illustrated Guide to Famous Places in the Province of Owari, vol. 2), 1880, (http://dl.ndl.go.jp/info:ndljp/pid/764881), p. 24.

3 Hiraga Gennai sensei kenshōkai (eds), *Hiraga Gennai zenshū, ge* (Collected works of Hiraga Gennai, vol. 2), Tōkyō: Hiraga Gennai sensei kenshōkai, 1934, pp. 1501, 1502.

4 Terakado Seiken et al., *Edo hanjōki* (Report on the prosperity of Edo), Edo, 1832–36.

These new investigations validated empirical analysis over the traditional reliance on textual authorities, and so, scholars begun to study materials for themselves.

Many societies and groups devoted to natural history were established, and there were *yakuhin-e* exhibitions held throughout the country. For example, the Shogunal Medical Academy in Edo held the first *yakuhin-e* in 1792 and every year thereafter until 1867, shortly before the end of the shogunate system.[5] *Yakuhin-e* exhibitions were usually open to the general public and free of charge. The Shogunal Medical Academy, allied with the lords and direct retainers of the shogun, established an association of natural history called the Shabenkai (lit. 'Association of the Red Rod") and this group met monthly to identify specimens and their taxonomy, and also to display rare finds.[6]

Several *yakuhin-e* exhibitions took place outside Edo, e.g. in Fukui, Wakayama, and Kumamoto.[7] In Ōsaka, Toda Gyokuzan (1696–1769) organized a *yakubutsukai* (also called *bussankai*) on four occasions beginning in 1760, and Iwanaga Buntei (1802–1866), a doctor and herbalist, held annual *yakuhin-e* exhibitions starting in 1835. Iwanaga Buntei produced and distributed advertisements for *yakuhin-e* exhibitions in 1760, and exhibits were gathered from Ōsaka, Kyō (the former name of Kyōto), Edo, Nagasaki, Akashi, the Sanuki region and others. A list of articles was published after the *yakuhin-e* exhibitions ended.[8] In Nagoya, Asai Shizan, a doctor to the lord of the feudal domain of Owari, organized annual *yakuhin-e* exhibitions at the domain's medical school from 1831. Mizutani Toyofumi (1779–1833), an herbalist, organized an herbalist group called *Shōhyaku-sha* in the feudal domain of Owari. Itō Keisuke (1803–1901), an herbalist and a member of this group, later studied under Philipp Franz von Siebold, during his first stay in Japan (1823–1829).

---

5 Isono Naohide, 'Yakuhin-kai/bussankai nenpyō (zōteiban)' (Exhibitions of natural products in the Edo period. A chronological table), in *Keiō gijuku daigaku Hiyoshi kiyō (shizen kagaku)*, vol. 29, 2001, pp 55–65.

6 For details see Federico Marcon, *The Knowledge of Nature and the Nature of Knowledge in Early Modern Japan*, Chicago: University of Chicago Press, 2015, pp. 189–191.

7 Isono, *The Knowledge of Nature*, p. 57, 61.

8 Kawasaki Eiko, 'Henshūsareru yakuhin-e no chi; 'Bunkairoku' to 'Shaben yoroku' wo chūshin ni' (The published knowledge of yakuhin-e: Focusing on 'bunkairoku' and 'shaben yoroku'), in *Hōsei daigaku daigakuin kiyō*, vol. 70, 2013, pp. 49–57.

3. FROM EDO TO TŌKYŌ: FROM *YAKUHIN-E* TO *HAKURANKAI*

The mission sent to the Exposition Universelle de Paris by the Shogunate in 1867 arrived there in March 1867. The special emissary, Tokugawa Akitake, was a younger brother of the 15th shogun, Tokugawa Yoshinobu. It was Japan's first official participation in an international exposition.[9] Kurimoto Joun (1822–1897), who was dispatched as a magistrate of foreign affairs, translated 'exposition' into Japanese as *hakurankai.* He also wrote 'Theory of the Museum' (*Hakubutsukan ron*) in 1875.[10] Alexander von Siebold, son of Philipp Franz von Siebold, was also part of this delegation and served as an interpreter. The Satsuma and Saga domains also participated independently in this exhibition. Saga domain, which was represented by Sano Tsunetami, participated because they wanted to purchase a warship and to exploit the overseas porcelain market. Satsuma domain exhibited its objects as a sovereign state.

After the Meiji Restoration in 1868, exhibitions were held under the name *hakurankai.* There were many natural products (taxidermied specimens, botanical and zoological specimens) on display at such *hakurankai* exhibitions, alongside antiquities, historical specimens, rarities, etc. Hence there was at that time a strong similarity and continuity between *yakuhin-e* exhibitions and the *hakurankai* exhibitions.

In 1872, the state's first formally titled *hakurankai* exhibition was held on the premises of the former Yushima seidō in Tōkyō at the Department of Natural History (*hakubutsu-kyoku*) of the Ministry of Education. The exhibition was held to prepare for participation in the international exposition in Vienna in 1873.

---

9 The first display of Japanese objects at an international exhibition in London in 1862 had coincided with the arrival in the city of a mission from Japan, the first Japanese mission to Europe after the opening of the country. The objects were the property of the first British Minister Sir Rutherford Alcock. There were 614 exhibits, ranging from ceramics and lacquerware to rain cowls made of bark and reeds. Fuchibe Tokuzō, a member of the mission, complained that the Japanese section looked just like an antique store, and that he could hardly bear to view the exhibits; see Fuchibe Tokuzō, 'Ō-kō nikki' (Journal of a trip to Europe) in Nihon-shi sekikyōkai (eds), *Kengai shisetsu nikki sanshū 3*, Tōkyō: Tōkyō daigaku shuppankai, 1971, p. 50. Fukuzawa Yūkichi who was also a member of the mission, after the trip translated the term 'museum' into Japanese as 'hakubutsukan'.

10 The text is included in Aoki Yutaka (ed.), *Meiji-ki hakubutsukan-gaku kihon bunken shūsei* (Collection of primary materials on museology in the Meiji period), Tōkyō: Yūzankaku, 2013.

On 14 January 1872, an edict was issued ordering prefectures to send their exhibits to the government. The government requested the submission of natural and manufactured products prior to the opening on 10 March. The exhibition was scheduled to run until the end of March, but it was extended throughout April due to popular demand. The total attendance was estimated to be more than 190,000.

After the exhibition ended, the majority of the items were sent to Vienna, with the remainder being exhibited at Yamashitamon-nai Museum. The museum had originally been a storehouse for the remains of the residence of the Satsuma domain in Edo. It was named Tōkyō Hakubutsukan in 1875, later becoming the Tōkyō National Museum.[11]

The first large-scale public *hakurankai* was organized in Kyōto in 1871 by three local businessmen and entrepreneurs: Mitsui Hachirōemon, head of the Mitsui family and its business, Ōno Zensuke, businessman and industrial pioneer, and Kumagai Naotaka, another prominent elite merchant from Kyōto. The exhibition was promoted as a means of expanding knowledge and craft techniques and of restoring the former capital Kyōto, since the organizers were concerned that the transfer of the new capital eastwards would cause their position to decline. The *hakurankai* exhibition was held at Nishi Honganji, one of the largest temples in Kyōto, from 10 October to 11 November 1871. Four days after the exhibition, the Kyōto Hakurankai Kaisha (Kyōto Exhibition Company) was established, and the company held the exhibition from the following year.[12]

For the new Meiji government, the *Weltaustellung* (World Exposition) in Vienna in 1873 was the first opportunity for Japan to officially participate in the international exposition since the Restoration. It meant that there was a strong desire on the part of the government to promote a new Japan to the world. Minerals were the main export items in the Edo period, but craft

---

[11] The British architect Josiah Conder completed the building of the museum in 1881. It was christened the Fine Arts Building for the Second National Exhibition. After the event collections and records were transferred from Yamashitamon-nai Museum, and it opened as the Imperial Museum in 1882.

[12] Maruyama Hiroshi, 'Meiji shoki no Kyōto hakurankai' (Exhibitions in Kyōto in the early Meiji period), in Yoshida Mitsukuni (ed.), *Bankoku hakurankai no kenkyū*, Kyōto: Shibunkaku shuppan, 1986, pp. 221–248; esp. 227, 229; Kudō Yasuko, 'Meiji shoki Kyōto no hakurankai to kankō' (A Study of exhibitions and tourism in the early Meiji era), in *Kōka Joshidaigaku kenkyū kiyō*, vol. 46, 2008, pp. 77–100.

products, such as porcelain and fans, had proved extremely popular at the previous international exposition (Paris 1867), the first in which Japan had officially participated. Such exhibits triggered the popularity of Japonisme, and Japanese crafts became so popular that the Japanese government decided to send a wide range of crafts items, such as porcelain (*Arita yaki*), fans (*sensu* and *uchiwa*), tea, *katagami* (paper stencil), etc., and these received much attention and critical acclaim. The exposition report noted that more than three thousand fans had been sold in one day during the exposition. A semi-governmental company named Kiryū Kōshō Kaisha was established during the exposition to sell Japanese exhibits.

After the Vienna exposition, Ōkubo Toshimichi, a member of the mission who went on to become the first Minister of Home Affairs, decided to hold a national industrial exhibition to encourage industry. It was Ōkubo himself, in his ministerial capacity, who opened the first of five national industrial exhibitions during the Meiji era on 21 August 1877 in Ueno Park in Tōkyō.

These national industrial exhibitions were held in Tōkyō, Kyōto, and Ōsaka. The first three exhibitions were held in Tōkyō (1877, 1881, 1890), the fourth was held in Kyōto (1895), and the fifth in Ōsaka (1903).[13] The products left over from the first national industrial exhibition were sold at a *kankōba* which opened in January 1878. A kind of bazaar, these *kankōba* were fixed location venues, which became extremely popular in the Meiji era for making direct purchases, and were the precursor of department stores in Japan.[14]

In the late 1880s, the Ministry of Foreign Affairs, the Ministry of Agriculture and Commerce, the Ministry of Education and a Business Corporation (which was led by the prominent

---

13 Kuni Takeyuki, *Hakurankai no jidai: Meiji seifu no hakurankai seisaku* (The time of exhibitions: The exhibition policy of the Meiji government), Tōkyō: Iwata shoin, 2005,

14 Suzuki Hideo, *Kankōba no kenkyū: Meiji bunka to no kakawari* (A study of kankōba and its relations to Meiji culture), Tōkyō: Sōeisha (Sanseidō shoten), 2001; Miyake Takuya, *Kindai Nihon 'chinretsujo' kenkyū* (Commercial museums in modern Japan), Kyōto: Shibunkaku shuppan, 2015, pp. 97–105. On the birth of department store in Japan see Hatsuda Tōru, *Hyakkaten no tanjō: Meiji Taishō Shōwa no toshi bunka wo enshutsu shita hyakkaten to kankōba no kindai-shi* (Birth of the department Store: Modern history of the department store and the arcade, which staged the urban culture of Meiji, Taishō and Shōwa), Tōkyō: Sanseidō, 1993.

entrepreneur Shibusawa Eiichi) established a commercial museum,[15] paving the way for the establishment of industrial promotion halls (*shōhin chinretsujo*) in the prefectures.[16] The industrial promotion halls functioned as showcases for local specialties and became symbolic venues for their region.

The Minister for Home Affairs, Ōkubo Toshimichi, insisted that the *hakurankai* exhibitions were events that facilitated the acquisition of new knowledge and technology. Over time, however, *hakurankai* exhibitions came to incorporate an element of entertainment value and were seen as a place for the public to amuse themselves and to enjoy viewing new technologies and appliances. For example, the fifth national industrial exhibition in Ōsaka in 1903 included an elevator and a water slide. *Yakuhin-e, hakurankai* and the other types of exhibitions attracted and fascinated people by exhibiting innovations in technology and knowledge, rarities, curiosities and exotics.

4. DISSEMINATION OF TECHNOLOGY AFTER THE INTERNATIONAL EXHIBITION IN VIENNA

Sano Tsunetami (1822–1902), who was the administrative vice-governor for the Vienna World Exhibition, insisted that artisans should be sent to Vienna to study their own field. Seventy members of the delegation sent to Vienna, including twenty-four artisans, would remain in Europe to study and bring back various technologies relevant to their respective fields.[17] After the exhibition, an official report was edited by Hirayama Narinobu (1854–1929), who was a secretary at the exhibition, and Tanaka Yoshio (1838–1916), a civil servant and member of the Museums Bureau. The last volume of the report, titled *The Dissemination of Technology*, describes how this learning process took place, and deals with various crafts, including sericulture, textile, silk reeling, dyeing, watch manufacture, glass manufacture, and porcelain manufacture.[18]

Kaijiro Notomi (1844–1918), later known as an industrial art pioneer, who had been sent to Vienna as a jury member in the

[15] Miyake, *Kindai Nihon 'chinretsujo' kenkyū*, p. 146

[16] Miyake, *Kindai Nihon 'chinretsujo' kenkyū*, p. 146

[17] Fujiwara Takao, *Meiji zenki Nihon no gijutsu denshū to iten: Uīn bankoku hakurankai no kenkyū* (Technical training and transfer in early Meiji Japan: Studies of the International Exhibition in Vienna), Tōkyō, Maruzen puranetto, 2016, pp. 100–105.

[18] Fujiwara, *Meiji zenki Nihon no gijutsu denshū to iten*, pp. 130–192.

porcelain section, found out about ceramic industries in Elbogen (Loket) in Bohemia and also visited the Sèvres porcelain factory in France, which was then one of the principal porcelain manufactories not only in France, but the whole of Europe.[19] After returning to Japan, he disseminated production methods for factories and supervised the *Onchi zuroku*, an 84-volume collection of craft design sketches for Japanese craft objects, which were to be exhibited in international and domestic exhibitions or exported to Western countries. These sketches were distributed around the Japanese archipelago. Exhibitions for which they were used include the 1876 Philadelphia Centennial International Exhibition, the 1877 National Exhibition, the 1878 Paris International Exposition and the 1881 Second National Exhibition.[20] Besides Kaijiro Notomi, other artisans who had learned western technology and acquired new knowledge in connection with the Vienna exhibition and their European travels, also disseminated their experiences. Some of them would even establish or teach at schools and factories.[21]

Alongside the national industrial expositions, the government promoted another type of exhibition: *kyōshinkai*, competitive exhibitions or prize shows. The aim of these exhibitions was to disseminate knowledge of techniques, educate local industry, and improve standards and competition among craftsman. They were held routinely in each prefecture and supported by the Ministry of Agriculture and Commerce,[22] and exhibited items borrowed by the ministry, and machines and local specialties owned by prefectures.[23] The difference between domestic industrial expositions and competitive exhibitions was that competitive exhibitions prioritized the improvement of local industry.

---

19 Fujiwara, *Meiji zenki Nihon no gijutsu denshū to iten*, pp. 180–184.

20 Tōkyō National Museum (ed.), *Meiji dezain no tanjō: chōsa kenkyū hōkokusho 'Onchi zuroku'* (The birth of Meiji design: Report of research on 'Onchi zuroku' (a collection of craft design sketches)), Tōkyō: Kokusho Kankōkai, 1997.

21 Tomoda Kiyohiko, 'Naimushō-ki nōsei jitsumu kanryō to kan-nō seisaku no tenkai' (Secretarial Bureau Drafts of Agricultural Policy Ministry and Agricultural Policy during the Period of the Naimusho (Home Ministry)), in *Nōson kenkyū*, vol. 106, 2008, pp.1–12.

22 Kiyokawa Yukihiko, 'Shokusan kōgyō seisaku to hakurankai/kyōshinkai no igi – sono fukyū sokushin kino no hyōka' (The economic significance of national industrial exhibitions and competitive shows in Meiji era – An evaluation from the viewpoint of technological diffusion), in *Keizai kenkyū* (Hitotsubashi University), vol. 39–4, 1988, pp. 340–359.

23 Kiyokawa, 'Shokusan kōgyō seisaku to hakurankai/kyōshinkai no igi', p. 343.

5. CONCLUSION

Exhibitions were one of the most useful media for disseminating new knowledge and technology in the eighteenth and nineteenth centuries in Japan. The innovations, rarities and curiosities on display awakened the inquisitiveness and stoked the imagination of those who attended. Seeing is believing and, as the public viewed the exhibits, they also learned about the objects in a spirit of joy and amusement.

During the Edo period, the *yakuhin-e* exhibitions played an important role not just in the development of natural history in Japan, but also in the improvement in learning styles. Scholars began to collect real objects and exchange knowledge through exhibits and conferences. The expansion of *yakuhin-e* exhibitions opened up new research methods to scholars, and their extended social networks and express messenger system facilitated the exchange of knowledge and objects for themselves.

After the Meiji Restoration, modernization was a national policy. Hence, the government sought to find a way of disseminating western knowledge and the latest technologies to the public. The government decided to participate in the International Exposition in Vienna in 1873 and to organize the National Industrial Exhibitions.

Learning, which is achieved through observation or practical experience, greatly increases the possibility of improveing one's technique through imitation, of being inspired, and of innovating and creating. Exhibitions can promote this type of learning through introducing new techniques and improving existing techniques through competitive participation. Exhibitions are an especially practical medium for the dissemination of knowledge and technology. This was understood by the Japanese government of the time and was the driving force behind the initiative to organize the national industrial exhibitions, competitive exhibitions, and commercial museums.

REFERENCES

Aoki Yutaka (ed.), *Meiji-ki hakubutsukan-gaku kihon bunken shūsei* (Collection of primary materials on museology in Meiji period), Tōkyō: Yūzankaku, 2013.

Barber, Lynn, *The Heyday of Natural History, 1820–1870*, London: Jonathan Cape, 1980.

Fuchibe Tokuzō, 'Ō-kō Nikki' (Journal of a trip to Europe), in Nihonshi sekikyōkai (ed.), *Kengai shisetsu nikki sanshū 3*, Tōkyō: Tōkyō daigaku shuppan-kai, 1971.

Fujiwara Takao, *Meiji zenki Nihon no gijutsu denshū to iten: Uīn bankoku hakurankai no kenkyū* (Technical training and transfer in early Meiji Japan: Studies of the International Exhibition in Vienna), Tōkyō, Maruzen puranetto, 2016.

Hatsuda Tōru, *Hyakkaten no tanjō: Meiji Taishō Shōwa no toshi bunka wo enshutsu shita hyakkaten to kankōba no kindai*-shi (Birth of the department store: Modern History of the department store and the arcade, which staged the urban culture of Meiji, Taishō and Shōwa), Tōkyō: Sanseidō, 1993.

Hiraga Gennai sensei kenshōkai (ed.), *Hiraga gennai zenshū, ge* (Collected works of Hiraga Gennai, vol. 2), Tōkyō: Hiraga Gennai sensei kenshō kai, 1934.

Imahashi Riko, *Edo no kachō-ga hakubutsu-gaku wo meguru bunka to sono hyōshō* (Birds and flowers: The representation of natural history during the Edo period), Tōkyō: Sukaidoa, 1996.

Isono Naohide, 'Yakuhin-kai/bussankai nenpyō (zōteiban)" (Exhibitions of natural products in the Edo period. A chronological table), in *Keiō gijuku daigaku Hiyoshi kiyō (Shizen kagaku)*, vol. 29, 2001, pp. 55–65.

Kawasaki Eiko, 'Henshūsareru yakuhin-e no chi; 'Bunkairoku' to 'Shaben yoroku' wo chūshin ni" (The published knowledge of *yakuhin-e*: Focusing on 'bunkairoku' and 'shaben yoroku', in *Hōsei daigaku daigakuin kiyō,* vol. 70, 2013, pp. 49–57.

Kudō Yasuko, 'Meiji shoki Kyōto no hakurankai to kankō" (A Study of exhibitions and tourism in the early Meiji era), in *Kōka Joshidaigaku kenkyū kiyō*, vol. 46, 2008, pp. 77–100.

Kuni Takeyuki, *Hakurankai no jidai: Meiji seifu no hakurankai seisaku* (The time of exhibitions: The exhibition policy of the Meiji government), Tōkyō: Iwata shoin, 2005.

Kiyokawa Yukihiko, 'Shokusan kōgyō seisaku to hakurankai/kyō shinkai no igi – sono fukyū sokushin kino no hyōka' (The economic significance of national industrial exhibitions and competitive shows in Meiji Era – An evaluation from the viewpoint of technological diffusion), in *Keizai Kenkyū* (Hitotsubashi University), vol. 39, no. 4, 1988, pp. 340–359.

Marcon, Federico, *The Knowledge of Nature and the Nature of Knowledge in Early Modern Japan,* Chicago: University of Chicago Press, 2015.

Maruyama Hiroshi, 'Meiji shoki no Kyōto hakurankai" (Exhibitions in Kyōto in the early Meiji period), in Yoshida Mitsukuni (ed.), *Bankoku hakurankai no kenkyū*, Kyōto: Shibunkaku shuppan, 1986, pp. 221–248.

Merrill, Lynn L., *The Romance of Victorian Natural History,* Oxford University Press, 1989.

Miyake Takuya, *Kindai Nihon 'chinretsujo' kenkyū* (Commercial museums in modern Japan), Kyoto: Shibunkaku shuppan, 2015.

Miyoshi Nobuhiro, *Nihon kōgyō kyōiku seiritsu-shi no kenkyū* (Studies on the formation of technical education in Japan), Tōkyō: Kazama shobō, 1979.

Nishimura Saburō, *Bunmei no naka no hakubutsu-gaku: Sei-Ō to Nihon* (Natural history in civilizations: The West and Japan), Tōkyō: Kinokuniya, 1999.

Suzuki Hideo, *Kankōba no kenkyū: Meiji bunka to no kakawari* (The study of *kankōba* and its relations to Meiji culture), Tōkyō: Sōeisha (Sanseidō shoten), 2001.

Terakado Seiken et al., *Edo hanjōki* (Report on the prosperity of Edo), Edo, 1832–36.

Tōkyō National Museum (ed.), *Meiji dezain no tanjō: chōsa kenkyū hōkokusho 'Onchi zuroku'* (The birth of Meiji design: Report of research on 'Onchi zuroku' (a collection of craft design sketches)), Tōkyō, Kokusho kankōkai, 1997.

Tomoda Kiyohiko, 'Naimushō-ki nōsei jitsumu kanryō to kan-nō seisaku no tenkai' (Secretarial bureau drafts of Agricultural Policy Ministry and Agricultural policy during the period of the Naimusho (Home Ministry)), in *Nōson kenkyū*, vol. 106, 2008, pp.1–12.

Uemura Shōji, 'Meiji shoki kōgaku kyōiku kikan no setsuritsu – Kōgaku-ryō ni tsuite' (The establishment of the Imperial College of Engineering in the early Meiji period), in *Shakai kagaku*, vol. 40, no. 3, 2010, pp. 21–47.

INTERNET SOURCES:

Owari meisho zue, zenhen 2 (Illustrated guide to famous places in the province of Owari, vol. 2), 1880, (http://dl.ndl.go.jp/info:ndljp/pid/764881).

4

Knowledge on Mining and Smelting and Its Dissemination in the Edo Period

Regine MATHIAS

1. INTRODUCTION: MINING IN JAPAN

ALTHOUGH JAPAN IS today widely considered to be lacking in natural mineral resources, the country did have rich ore deposits in historical times. Gold, silver and copper deposits were concentrated in the Tōhoku region and the Kansai area. Gold deposits were also found on the island of Sado, in the mountains of the Kantō region, on the Izu peninsula, and in Satsuma (present-day Kagoshima prefecture), and copper was mined in many parts of Japan.

During the Edo period (1600–1868), among the most important mines were the silver mines of Iwami/Ōmori[1] (Shimane), Ikuno (Hyōgo) and Innai (Akita), the gold mines of Sado (Niigata), the Tōhoku area, the prefectures of Yamanashi and Kagoshima, as well as the copper mines of Ani (Akita) and Besshi (Ehime) (see Fig. 1).

At the end of the Edo period, iron was found mainly in the form of sand iron in many parts of the country, but especially in western Honshū, in the prefectures of Shimane and Tottori. Deposits of lead and tin, which were important for processing gold, silver and copper, were relatively small, and so some of these ores had to be imported.

[1] The town consisted of two parts: Iwami, the mining town, and Ōmori, the administrative center. While Ōmori is the official name used today, Iwami is still used in many publications, especially concerning the silver mine.

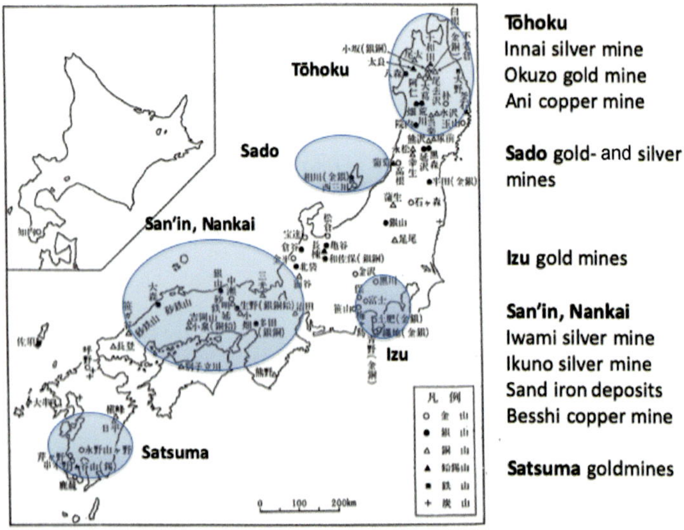

Fig. 1: Major mining areas of precious metals in Japan.
Source: Based on Kobata, *Nihon kōzan shi no kenkyū*, p. 4.

This chapter focuses on gold, silver and copper mining and the knowledge and technology necessary for successfully carrying out mining of these minerals. Iron was the most consumed metal, and thus the most intensively mined. Despite its importance, techniques for extracting and processing its raw material, sand iron, have hardly evolved over the centuries. In contrast, the development and dissemination of new knowledge played an important role in the increase in mining of precious metals during the late sixteenth and early seventeenth centuries.

Knowledge related to mining roughly covers the following areas:

- Finding the deposits: Prospecting, surveying
- Work underground: Heading, drifting, mining, ore extraction, transport, drainage and ventilation
- Further processing: Dressing – smelting – refining

Until the end of the Edo period, knowledge in all of these areas was essentially empirical, and was sometimes supplemented with cosmological knowledge for activities such as prospecting.

2. DEVELOPMENT OF MINING AND MINING TECHNOLOGY BEFORE THE SIXTEENTH CENTURY

Metals were first used in Japan during the Yayoi period (ca 400 BCE to 300 CE), when metal tools and knowledge about metal processing were gradually introduced from China and Korea. The development of ore mining probably began several centuries later in the late Kofun period (ca 300 to 710 BCE).[2] It can be traced back to at least the seventh and early eighth centuries through written reports in the ancient sources *Kojiki* (712), *Nihon shoki* (720), *Shoku Nihongi* (797), as well as in individual topographies such as *Izumo fudoki* (733) and *Harima fudoki* (713–15), among others.

The seventh and eighth centuries were a crucial period for establishing the foundations of mining knowledge in Japan. The *Nihon shoki* recorded the first discovery of silver on the island of Tsushima, situated between Kyūshū and the Korean peninsula, in 674. The silver was subsequently paid as a tribute to the imperial court. On Tsushima, silver is found in galenite (a silver-containing lead sulphide). To extract the silver, one must already know how to separate silver from the lead. It is assumed that at the time people used a kind of cupellation process,[3] and that knowledge of this procedure likely came from China.[4] Evidence for how Japanese acquired this knowledge is scarce, but knowledge about cupellation for obtaining silver from sulphide ores was described in Chinese texts in the Tang (618–907) and Song periods (960–1279),[5] and could have reached Tsushima via the Korean peninsula in the seventh century.

The *Tsushima no kuni kōginki*, a short text believed to have been written in the late ninth century, is an early source describing

2 The development of ore mining and metal processing in Japan differs in two respects from the usual sequence. First, the use and processing of bronze and iron began almost simultaneously. Second, ore mining can only be proven much later than the processing of metals. See Murakami Ryū, *Kin gin dō no Nihon-shi* (The Japanese history of gold, silver and copper), Tōkyō: Iwanami shoten, 2007, pp. v–vi.

3 Cupellation is a refining process to separate noble metals like gold and silver, from base metals, like lead or copper, present in the ore (see also below).

4 Nakanishi Tetsuya, Izawa Eiji, 'Evolution of Silver-smelting Technology on Japan in the Middle of the Sixteenth Century', in *ISIJ International*, vol. 54 (2014), no. 5, pp. 1093–1097, p. 1093.

5 Peter J. Golas (and Joseph Needham), *Chemistry and Chemical Technology*, part XIII: *Mining*, Cambridge: Cambridge University Press, 1999, pp. 132–133.

silver mining and processing on Tsushima.[6] It mentions the use of adits, which could reach a length of over a hundred metres, for silver mining. These adits were so deep that torches had to be used to work on them even during broad daylight. The extracted ore was smelted in a simple hearth, evidently using natural ventilation and no charcoal.[7] It is presumed that digging started at the outcrop of the ore and followed the vein to a certain depth.

Besides silver and gold, so-called precious metals, including copper, and copper alloys such as bronze were also important. The *Shoku Nihongi* mentions that, during the rule of Emperor Mommu (r. 697–707), copper found in the provinces of Inaba (present-day Tottori prefecture) and Suō (present-day Yamaguchi prefecture) was presented to the court. Emperor Mommu was an ardent promoter of mining and sent expeditions to explore new deposits. He also included the first regulations for mining activities in the Taihō Code of 701. Copper was highly valued, and when a large copper deposit was discovered in 707, Mommu's successor proclaimed the start of a new era named Wadō (lit. Japanese copper, 708–715), and new copper coins were cast and brought into circulation. The development of mining, refining and casting reached a first peak in the middle of the eighth century when a large bronze statue, the Great Buddha of the Tōdaiji temple in Nara, was cast and gilded. The nearly 500 tons of copper used for this statue came from several copper mines in the western part of Honshū.[8]

The mining methods used for copper must have been similar to those used for silver. Little is known about the smelting processes, but archaeological findings hint at two types of furnaces for smelting and refining. Copper oxide was probably melted in a kind of crucible furnace (*rutsubo ro*), similar to those furnaces used for casting bronze. Copper sulphide, on the other hand, appears to have been processed in an earth furnace to avoid chemical reactions

---

6 Azuma Kiyoshi, 'Tsushima no kuni kōginki to sono seirenhō' (The 'Tsushima kuni kōginki' and its smelting method), in: *Nihon kōgyō kaishi* 91 (1051), 1975–9, pp. 607–609. The short text of the *Tsushima no kuni kōginki* (Report on silver tributes from the province of Tsushima) is reprinted in Saigusa Hiroto (ed.), *Nihon kagaku koten zensho*, vol. 9, part 3, *Sangyō gijutsu hen. Saikō yakin* 1, pp. 223–229. Author (Ōe no Koretoki ?) and date are not clear, but there is evidence that points to the ninth century.

7 Azuma describes why, even without the use of charcoal and bellows, the process described in the text could have produced silver.

8 Bruno Lewin, (ed.), *Kodō zuroku. Illustrierte Abhandlung über die Verhuettung des Kupfers*, 1801, Bochum: Deutsches Bergbau-Museum, 1984, p. 7

with the wall of the crucible furnace.[9] It is assumed that for the large quantities needed for projects such as the Great Buddha in Nara or large temple bells more complex procedures, such as roasting or kernel roasting, might have been used to separate copper from compounds.

During the Nara (710–794) and early Heian periods (794–1185), the dissemination of mining knowledge appears to have been actively promoted by the authorities. It is recorded that in 878, an imperial order issued by the Dazaifu, the imperial office in Kyūshū, the province of Buzen (today parts of Fukuoka and Ōita prefectures) provided one hundred men in statute-labour service (*yōfu*) to dig for copper in the Kiku-*gun* (north eastern part of Buzen). In 885, the Naganobori mine in the Suō province, one of the oldest copper mines in Japan, was ordered to send one master craftsman for copper mining (*dō-te*) and one miner (*hori-ana-te*) to Buzen to the Head of the Copper Office (*saidō-shi*), in order to help the people there, who did not have any experience of copper mining.[10]

According to the *Nihon shoki*, the first gold in Japan was discovered in 749 in the Tōhoku region. Until the end of the sixteenth century, gold was mined in Japan primarily as placer gold (alluvial gold). The use of gold to gild statues like the large Buddha in Nara or the famous Golden Hall (Konjikidō) of the Chūzonji temple in Hiraizumi, erected in 1124, obviously contributed to Japan's reputation as 'Zipangu, the gold country', which gained currency through Marco Polo's travel accounts after his journey to China in the late thirteenth century. Over time, not only was gold washed in rivers, but people actively tore down the slopes of the mountains near rivers, and built ever more complex systems of ponds, trenches and dams to regulate, flood, and drain the rivers according to the needs of the gold miners.[11]

It can be said that from the end of the seventh and beginning of the eighth century the foundations of mining knowledge had been laid and were actively spread by government officials and 'specialists' in Japan. This knowledge came from China and/or the Korean peninsula and concerned the mining and processing of gold, silver and copper as well as lead and other minerals like

---

[9] Kanzaki Masaru, *Yakin kōkogaku gairon* (Outline of the archeology of metallurgy), Tōkyō: Yūsankaku, 2006, pp. 75–77

[10] Kobata Atsushi, *Nihon kōzan shi no kenkyū* (Studies in Japanese mining history), Tōkyō: Iwanami shoten, 1968, p. 50.

[11] Kobata, *Nihon kōzan shi*, pp. 45, 47–48.

mercury. However, this knowledge remained largely unchanged until the sixteenth century. In other words, although processes were refined and became more elaborate, no new technical knowledge, methods or tools were developed except where these related to construction (e.g. dam construction).

3. INCREASE IN MINING THROUGH THE INTRODUCTION OF NEW TECHNICAL KNOWLEDGE

The period of active promotion of mining and mining knowledge in ancient Japan was followed by a long period of stagnation or very slow development. This period, which mining historian Murakami Ryū describes as phase III in the development of mining scheme (see Fig. 2), lasted almost 800 years and only ended with the introduction of a new silver refining process in the Iwami silver mine in the 1530s. Phases IV and V, which follow phase III, are of particular interest for the development of mining knowledge in the Edo period and will provide the timeframe for the following remarks.[12]

In the sixteenth and seventeenth centuries, mining in Japan experienced an epochal boom. Costly wars in the late-fifteenth and sixteenth centuries, as well as the increasing foreign trade and global demand for silver (and gold), stimulated the search for precious metals and copper in Japan. Almost all major ore mines of the Edo period trace their origins back to this period. The boom was stimulated by new knowledge, which led to technical improvements in surveying, tunnel driving and smelting.[13]

• I Beginnings	草創	Yayoi period
• II Taking roots	定着	538 – 752
• III Groping	模索	752 – 1526
• IV Development	発展	1526 – 1596
• V Maturing	熟成	1596 – 1736
• VI Mature state	爛熟	1736 – 1853
• VII Reviving	再生	since 1853

Fig. 2: Stages of ore mining in Japan.
Source: Murakami, *Kin gin dō no Nihonshi*, p. vii.

[12] Murakami, *Kin gin dō no Nihonshi*, p. vii.
[13] Kobata, *Nihon kōzan shi*, p. 51

Major improvements resulted from the introduction of new technologies for dressing, smelting and refining ores. The new knowledge that was generated in the sixteenth and seventeenth centuries led to a significant increase in mining production. Among the new, or at least greatly improved, technologies were a much more efficient cupellation process (*haifuki*) for the extraction of silver, a process called Western smelting (*nanbanfuki*) for the separation of silver and copper, and a process using mercury to extract silver (and gold), called amalgamation (*suigin nagashi*). All technologies were based on new knowledge imported from China and Korea, from Europe, and from Mexico, respectively. In the last of these cases, the process of amalgamation was likely the same as that used by the Spanish conquerors.

The increase in mining in the sixteenth century started with a silver boom, triggered by the introduction of an enhanced *haifuki* cupellation technology to the Iwami silver mine in Ōmori (present Shimane prefecture). According to records, in 1526 a copper merchant from Hakata in Northern Kyūshū by the name of Kamiya Jutei obtained permission to exploit the deposits of the Iwami mine. His place of origin is important because at that time the port-town of Hakata had close relations with Korea, where silver mining was quite advanced. The ore was probably transported to Hakata initially where it underwent further processing. However, in 1533, Kamiya Jutei brought two smelting specialists, called Sōtan and Keishu, from his native Hakata to Iwami. They are said to have been of Korean origin. With their help, the new smelting process was successfully introduced to Iwami, and the production of silver started.[14]

In the process used in Iwami, silver-containing ores were melted with the addition of lead and manganese (as flux) (*subuki*). Impurities that rose to the surface were removed, creating an alloy of silver and lead. The lead-silver alloy was then separated by cupellation (*haifuki*). In this technique, the lead-silver alloy was melted again at a high temperature in a furnace lined with bone ash, while oxygen was added with a bellows. The lead oxidized

---

14 Nakano, *Ginzan shakai no kaimei*; p. ii; Kobata, *Nihon kōzan shi*, p. 51. For details concerning Kamiya Jutei and his background see Yoshioka Toya, *Iwami ginzan wo yomu. Kozu, emaki, kyūki, sekishūgin* (Reading Iwami silver mine. Old maps, picture scrolls, ancient reports, silver from Sekishū), Izumo: Hōkōsha, 2017, ch. 3. Because of its pioneering role in the development of early modern mining, Iwami was declared a World Heritage Site in 2007.

and bonded with the ash, while the silver remained on the ash. Due to this new technology, the amount of silver obtained at the Iwami mine, as well as its degree of purity, greatly increased.

Archaeological finds from Iwami indicate that a small iron vessel (23 cm diameter, 8 cm deep) was initially used for the

Fig. 3: *Haifuki* cupellation furnace.

This scene of a Sado mining scroll shows a series of cupellation furnaces.

Source: Picture scroll 23 *Ken rekihaku Sadokoku kinginzan shikioka kasegikata zu*, part 12a. Courtesy of Niigata Kenritsu Rekishi Hakubutsukan. Each furnace is operated by two people: the cupellation master and the bellows operator. The person sitting at the rear is the supervisor. Each furnace comprises a simple box bellows on the left and the furnace, a simple whole in the ground lined with ashes, in the middle. The material to be melted is placed between burning charcoal. During the process the fire is extinguished several times with water (to the right of the furnace) and the slag removed. The third furnace is not used and shows only the box bellows and the water bowl.

cupellation.[15] Later, a basic smelting furnace was developed by simply digging a circular cavity into the ground and lining it with a mixture of clay and charcoal dust (Fig. 3). The furnace had a hood to carry off the fumes and gases. The blast was produced by a wooden bellows construction. This kind of smelting furnace was used over the entire Edo period for various kinds of smelting processes, so that Western engineers who came to Japan spoke of a universal furnace.[16]

Silver from Iwami, and later from other Japanese mines, became an important export commodity. As early as in 1540, Korean authorities complained in the 'Annals of King Jungjong' (jp. *Chūsō jitsuroku*) about Japanese silver being brought in increasing quantities to Korea and taken via Korea to China, and stated that '[...] these Japanese learned the art of silver production with the help of a traitor from our country.'[17] In the years to come, Japan became an important player in the worldwide silver market, in which silver became a much sought-after commodity, which ultimately flowed primarily to China. News of the silver mine in Iwami spread to Europe, where Iwami was described as a place with *argenti fodinae* (silver mines) on several old maps (Fig. 4).

The new refining method spread all over the country from Iwami. In 1542 the Ikuno silver mine (present-day Hyōgo prefecture) was developed. Ikuno carried on the cupellation process used in Iwami and later became one of the most important sites for the production of silver for the evolving central power.[18]

Another new refining technology was the *nanbanfuki* or *nanban shibori* (Barbarian or Western smelting) process, a liquation process, similar to the Seiger process in Europe. It was mainly applied in copper production, but was also used to extract silver from argentiferous copper. According to the *Kodō zuroku* (Illustrated Book on the Smelting of Copper), a detailed and illustrated account of the production of copper in the Sumitomo

---

[15] For a picture and a description see Ōsaka Museum of History (ed.), *Tokubetsu ten: Yomigaeru akagane. Nanbanfuki to Sumitomo dōfukisho* (Special Exhibition: Rethinking copper. The *nanbanfuki* method and the Sumitomo copper refinery), Ōsaka Museum of History, 2003, p. 35.

[16] See e.g. Curt Netto, 'Ueber japanisches Berg- und Huetten-Wesen' (Japanese mining and smelting), in *Mittheilungen der Deutschen Gesellschaft für Natur- und Völkerkunde Ostasiens*, vol. II, no. 11–20, 1876–1880, pp. 365–409, esp. p. 377–378.

[17] Cited by Kobata, *Nihon kōzan shi no kenkyū*, p. 51.

[18] Kobata, *Nihon kōzan shi no kenkyū*, p. 51.

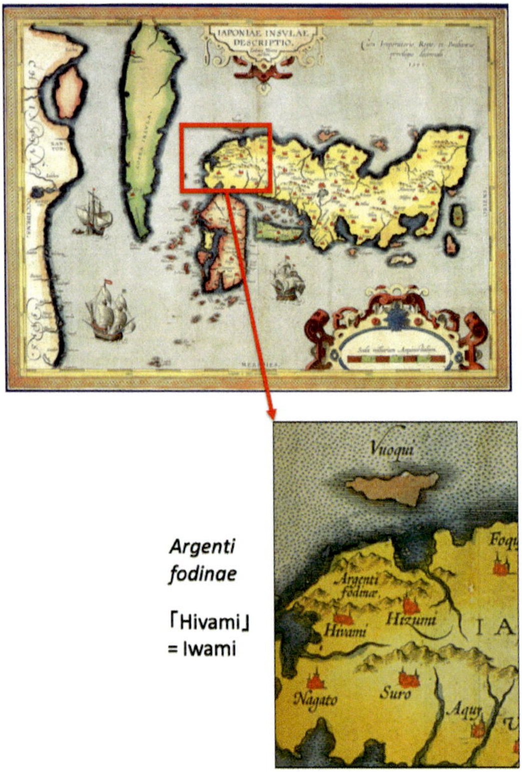

Fig. 4: Early Map of Japan showing Iwami silver mines.
This map was drawn by Luiz Teixeira, and published in the atlas by Abraham Ortelius, "Theatrum Orbis Terrarum", Antwerpen 1595. It shows Iwami (spelled as Ivami) as a place with silver mines.
Source: Walter, Lutz (ed.), *Japan mit den Augen des Westens gesehen. Gedruckte europaeische Landkarten vom fruehen 16. bis zum 19. Jahrhundert*, Muenchen, New York: Prestel 1993, Fig. 19.

Besshi copper mine (present-day Ehime prefecture) believed to have been printed in 1801, this procedure had not previously been possible in Japan. The *Kodō zuroku* also recorded that, in 1591, a member of the Sumitomo family, Sumitomo Jusai, heard about this new method of extracting silver from copper from a foreign merchant in the port city of Sakai.[19] This is often seen as

[19] The original by Masuda Tsuna, with illustrations by Niwa Motokuni Tōkei, was presumably printed in 1801. I refer to the facsimile edition with English

the beginning of the *nanban-shibori* process in Japan, but actually the origin of the new technology and how it made its way to Japan are not entirely clear. There can, however, be no doubt that the new technology played a crucial role, especially from the late seventeenth century, when copper replaced silver as an important export commodity.

The new technology needed a new type of furnace, which differed significantly from the simple furnaces used previously (Fig. 5). Silver-containing copper, to which lead has been added, is put into this furnace and heated until it is not quite liquefied. The lead

Fig. 5: *Nanban shibori* (Western Smelting)
Furnace for the separation of copper and silver-containing lead. In this kind of Seiger process alloyed copper is heated until the lead liquefies and runs off, while the copper remains in the furnace. Source: Smith, *Kodō zuroku,* plate XI, p. 49.

translation: Cyril Stanley Smith (ed.), *Kodō zuroku* (Illustrated Book on the Smelting of Copper), Norwalk, 1983, pp. 84–85.

will liquefy first and start to run out, carrying the silver with it. Therefore, a more elevated furnace is necessary. In contrast to the European Seiger process, in which the lead drains out naturally under gravity, in Japan the molten lead is squeezed out, and this results in a greater yield of silver, as well as a higher purity of copper. The remaining copper is then removed, and the silver lead alloy separated by cupellation.[20]

Knowledge of this new technology rapidly spread to silver and copper mines throughout the country.[21] In 1700, the Tokugawa *bakufu* established a Copper Agency (*dōza*) in Ōsaka and commissioned the Sumitomo family to manage copper refining in the city's various smelters. To ensure consistent quality, the *nanban-shibori* process was standardised across all the city's copper smelters.[22]

Yet another technology, an amalgamation process using mercury, is said to have been applied in Sado. However, it was used for only a very short time, and was soon abandoned. Knowledge of this process is still scarce, and will not be treated in further detail here.[23]

The introduction of new smelting technologies from the 1530s led to the emergence of various kinds of increasingly complex smelting processes, which enabled Japanese miners to exploit a broader variety of ore resources and extract precious metals in a more efficient way.

Another area of new knowledge, which greatly promoted the mining of precious metals in Japan, was civil engineering. In the second half of the sixteenth century, surveying technology (*sokuryō gijutsu*) experienced a real boost. Such technology was developed through the construction of numerous castles and irrigation systems, through land reclamation projects, and through the development of military science associated with the use of guns and cannons. In addition, the dissemination of nautical knowledge necessary for seafaring, and growing interest in mining and

---

20 Imai Noriko, 'Copper in Edo-Period Japan', in Keiko Nagase-Reimer (ed.), *Copper in the Early Sino-Japanese Trade*, Leiden, Boston: Brill, 2016, pp. 10–31, p. 15; Cyril Stanley Smith, 'Introduction', in *Kodō zuroku*, pp. 9–25, here pp. 17–23.

21 Kobata, *Nihon kōzan shi no kenkyū*, pp. 63–64.

22 Imai, 'Copper in Edo-Period Japan', pp. 17–18.

23 For more details see Tanaka Keiichi, 'Edo shoki amarugamu-hō no dōnyū to Ieyasu no bōeki seisaku' (The Introduction of the amalgamation method in the early Edo period and Ieyasu's trade policy), in: *Nihon rekishi* 501 (1990.2), pp. 71–84.

precious metals led to a leapfrogging development in surveying technology. There emerged various schools which refined traditional Chinese *Feng shui* (geomantic determination) methods or incorporated new methods of the Portuguese and Spanish (*nanban ryū*) or of the Dutch (*kōmō ryū*) into their practice.[24]

One of the centres for the development of such knowledge was the province of Kai (present-day Yamanashi prefecture). In Kai there were many rivers with strong currents, the regulation of which required knowledge about the construction of ponds and dams, the handling of water and rocks, as well as of surveying. Knowledge accumulated by the Kai school of civil engineering (*kōshū ryū*) was decisive in the development of mines all over Japan. Therefore, the province of Kai, along with the Iwami silver mine, are regarded as the two most important hubs for the generation of new mining knowledge.[25]

This leading role applies especially for the surveying processes. Already in the Keichō period (1596–1615) there were specialists, i.e. surveying technicians (*furigane-shi, sunpo*), who were responsible for tunnel digging in the mines. Instead of following the ore vein from the outcrop at the top of the mountain to ever-greater depths, tunnels were increasingly driven horizontally into the mountain from the slope at right angles to the presumed course of the ore vein. Once reached, the ore was dug upwards and downwards. This method was instrumental in enabling the exploitation of deeper ore veins and, according to some sources, likely developed under the influence of Western ideas.[26]

The *furigane-shi* surveyors were in charge of organising and directing the excavation.[27] To be successful, they had to be able to predict the course of the ore vein, and did so on the basis of accumulated empirical knowledge, as well as because of improved surveying methods. Various maps, which survived in Sado and other mines, showing the assumed course of the ore veins, are evidence of a more systematic approach to tunnel drilling (Fig. 6).

24 Suzuki Kazuyoshi and Tanabe Yoshikazu, 'Edo shoki no hōi oyobi kakudo no gainen kara mita sokuryōjutsu no keisei nit suite no ichi kōsatsu' (The concepts of 'angle' and 'direction' in schools of surveying method in the early Edo period), in: *Bulletin of the Museum of Natural Science,* Ser. E, 32, 2009, pp. 41–49, (http://www.kahaku.go.jp/research/publication/sci_engineer/v32.html, accessed 10.09.2018), p. 41–42.

25 Kobata, *Nihon kōzan shi no kenkyū*, p. 52.

26 Nakano, *Ginzan shakai no kaimei*, p. 23.

27 Kobata, *Nihon kōzan shi no kenkyū*, p. 53.

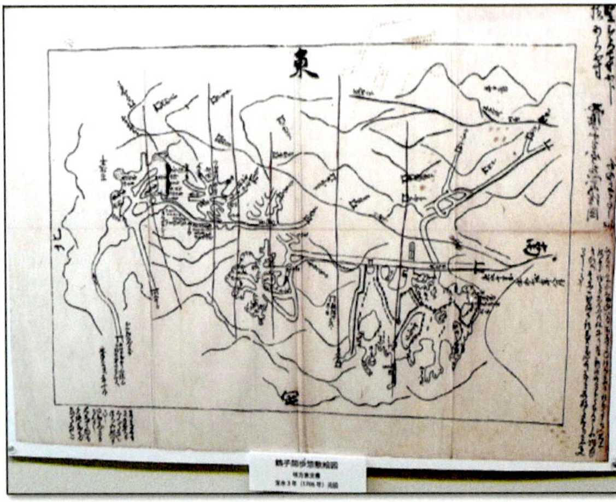

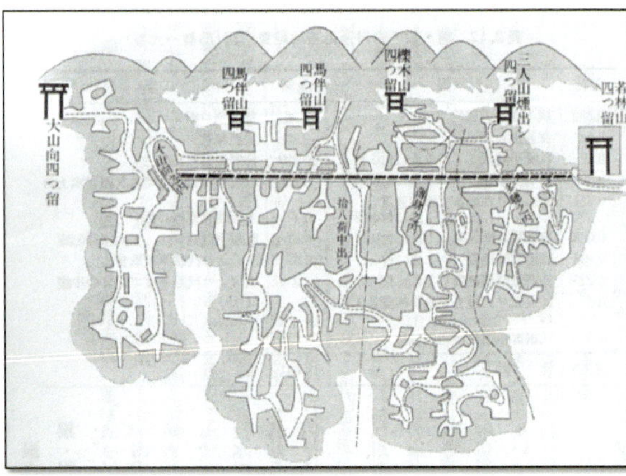

Fig. 6: Drawings of the adits and tunnels of two mines
The drawing above shows the Tsurushi silver mine in Sado around 1700.
Source: Drawing from 1706, Sado Museum. Courtesy of the Mikata family.
The drawing below depicts the main adit and various drifts in the Wakabashi Mine in Ikuno. There are no vertical shafts.
Source: Kobata, *Nihon kōzan shi*, p. 246, fig. 2.3.

Knowledge of civil engineering and surveying was particularly important for solving the water problem in the mines, and for ventilation. The building of long drainage tunnels had started in the early seventeenth century. One of the earliest books on mining, the *Kōzan shihō yōroku* (A Digest of Cherished Treasures on Mining), compiled by Kurosawa Motoshige, the head of the mining office (*kaneyama bugyō*) of Akita domain in 1691, already describes the methods for driving tunnels and adits into the mountains.[28]

At almost the same time, a 915 meter-long drainage tunnel was driven across a mountain in Sado within only six years, from 1691 until 1696, under the direction of a surveyor named Shizuno Yōemon. Yōemon had the tunnel advanced from three points in two directions each, i.e. six different sections were dug almost simultaneously. At the starting points, vertical shafts were sunk to the level of the tunnel. The measurements were so accurate that the excavations met with only minimal deviations. The Minamisawa Daisosui

Fig. 7: Minamisawa Drainage Tunnel
The longest of several drainage tunnels build in Sado during the seventeenth century. Built in a record time of six years, the tunnel reached a length of 900 meters, and was later extended to about 1000 meters. The cost was 1.4 kg of silver. The tunnel replaced 40 drainage pumps. Fumoto, *Sado kinginzan shiwa*, pp. 185–187. (Photo Regine Mathias).

[28] Kurosawa Motoshige, *Kōzan shihō yōroku* (A digest of cherished treasures on mining), in Saigusa Hiroto (ed.), *Nihon kagaku koten zensho*, vol. 10, part 3, *Sangyō gijutsu hen. Saikō, yakin 2*, Tōkyō, Ōsaka: Asahi Shinbunsha, pp. 1–75.

tunnel (Large Minamisawa Drainage Tunnel) still exists today and, at 2.40 metres high and 0.91–1.82 metres wide, it is an impressive example of the high level of surveying technology (Fig. 7).

Crucial for the success of such endeavours were surveying tools such as the compass. Magnetized needles had already been used in China since ancient times, when they were used mainly for *Feng shui*. Knowledge of how to use of the magnetic compass for this purpose (geomantic compass; Ch. *luopan, luoging*) and of techniques relating to magnetic needles had already reached mediaeval Japan. In addition, the plane table surveying method was developed in Japan.[29] Some of the surveying instruments used by Yōemon have survived. A magnetic compass (*rashinban*) and an inclinometer (*keishaki*) attributed to him are kept at the University of Tokyo. Another compass, which also dates back to his time, is preserved at Sado. In both compasses, the circumferences were divided into 480 degrees, which allowed a considerably precise measurement.[30]

Drainage tunnels (water draining adits) were immensely costly and were thus used only in exceptional cases. The usual form of drainage remained scooping up water with buckets, sometimes by using pulleys, or with pumps. At the beginning of the seventeenth century, piston pumps made of wood or bamboo, called *supontoi*, were already being used in many of the larger mines. According to the *Sado Nendaiki* (Chronicals of Sado), these were used because some of the miners already had knowledge of such pumps. In addition to these simple pumps, there were experiments with more advanced technologies at mines in Sado. A famous example is a type of Archimedean screw, the *suishōrin*, which was used for drainage around the middle of the seventeenth century (Fig. 8).[31]

Accounts of the origin of the *suishōrin* (in Japan) differ. According to some sources, a blind scholar from Ōsaka named Suigaku Sōhō was called to Sado in 1637, where he built a pump called *tatsutoi* (dragon pump). His pump outperformed the *supontoi* piston pumps and was used in Sado thereafter. Other sources mention that he came to Sado in 1653 and constructed an adapted Archimedean screw made of wood, which was called *suishorin*. The difference between the two is not clear, and some scholars

29 Suzuki/Tanabe, 'Edo shoki no hōi oyobi kakudo', p. 41.

30 For details and pictures see Nagase-Reimer, 'Water Drainage in the Mines in Tokugawa Japan', p. 30–32, and Suzuki/Tanabe, 'Edo shoki no hōi oyobi kakudo', pp. 45–46.

31 For details see Nagase-Reimer, 'Water Drainage in the Mines in Tokugawa Japan', pp. 32–34.

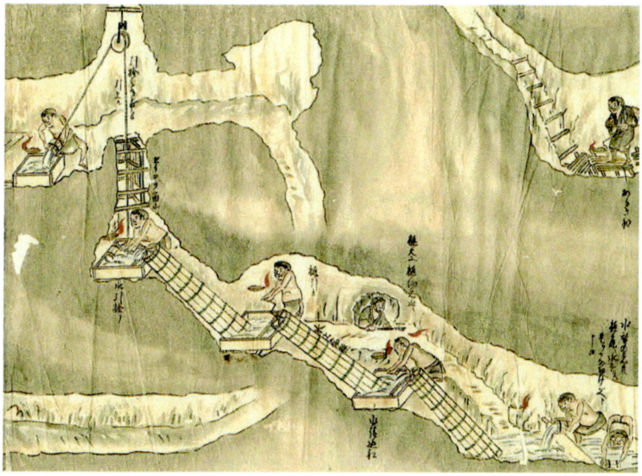

Fig. 8a: *Suishōrin* (Archimedean screw pump) in Sado.
As this picture of a mining scroll shows, several pumps were often connected in series in order to overcome the differences in altitude. The water was pumped up many metres, where it could then flow off.
Source: Picture scroll *Sado-shi Sado ginzan Ōji no kakō emaki*, Courtesy of Sado Shiritsu Sado Hakubutsukan.

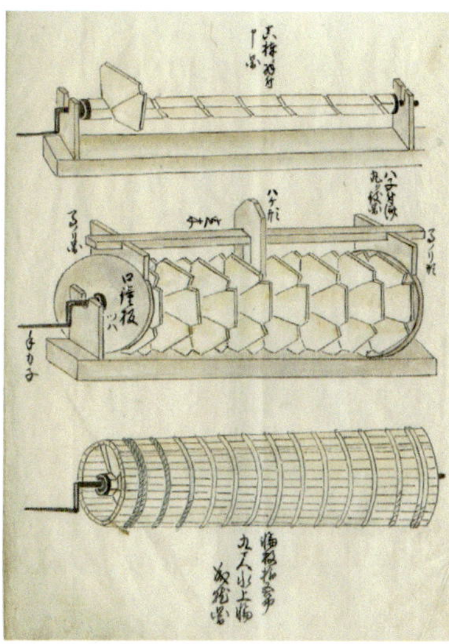

Fig. 8b: Construction of a *suishōrin*.
The drawing shows the central axis, the inner spiral and outer shell.
Source: *Kinginzan taigaisho* (General Information on Mining), date unknown. Courtesy of Sado Shiritsu Sado Hakubutsukan.

Fig. 8c: *Suishōrin*
This kind of Archimedean screw was used in Sado from 1653 to1679. The specimen shown here in the Aikawa Folk Museum of Sado only has its central axis and outer shell. The wooden blades forming the inner spiral are lacking (Photo Regine Mathias).

think that they were alike.[32] Suigaku Sōhō might have obtained his knowledge through a Chinese book, *Taixi shuifa* (Far Western Water Technology), written in 1612 by a Jesuit scholar from Italy and a Chinese scholar, in which an Archimedean screw was described.[33] The *suishorin* was a cylindrical device with a length of about 270 cm and a diameter between 30 and 40 cm, consisting of seven or eight bound hubs. Inside its centre shaft is a continual spiral screw. The whole device is turned with the crank handle and continuously lifts up water.[34]

In the years leading up to 1679 several hundred of these *suishorin* are said to have been built and used in Sado. To bridge the

32 Fumoto Saburō, *Sado kinginzan shiwa* (Historical Stories of the Sado Gold and Silver Mine), Tōkyō: Mitusbishi kinzoku kōgyō, 1973, pp. 133–138, esp. pp. 136–137. See also Nagase-Reimer, 'Water Drainage in the Mines in Tokugawa Japan', pp. 32–34.

33 See Nagase-Reimer, 'Water Drainage in the Mines in Tokugawa Japan', pp. 33. It is interesting to note that, according to Peter Golas, the Archimedean screw and some other pumps were not used in Chinese mines. See Peter J. Golas, Joseph Needham, *Science and Civilization in China,* vol. 5, *Chemistry and Chemical Technology,* part XIII: *Mining,* Cambridge: Cambridge University Press, 1999, p. 344. Golas discusses the use or absence of various forms of pumps in Chinese mining, pp. 341–349.

34 Nagase-Reimer, 'Water Drainage in the Mines in Tokugawa Japan', pp. 32.

height gap, several *suishorin* were often connected in series. They proved to be highly efficient, but susceptible to damage, particularly from water infiltration, and were exceedingly costly.[35]

More than a hundred years later, an attempt was made to solve the still-existing water problem with a Dutch-style brand spuijt (*furanka supoi,* a syringe pump), which the Dutch had brought to Japan in 1782 (Fig. 9). Several Japanese watchmakers seem to have copied this syringe pump, and seven such pumps were installed in Sado. Even though they proved to be highly efficient and cost-effective, they often broke down, and were not used in other mines. Even in Sado they went out of use after some time.[36] Instead, drainage with buckets and pulley blocks once again became the preferred drainage method, especially because cheap labour supplied by the central government was available.

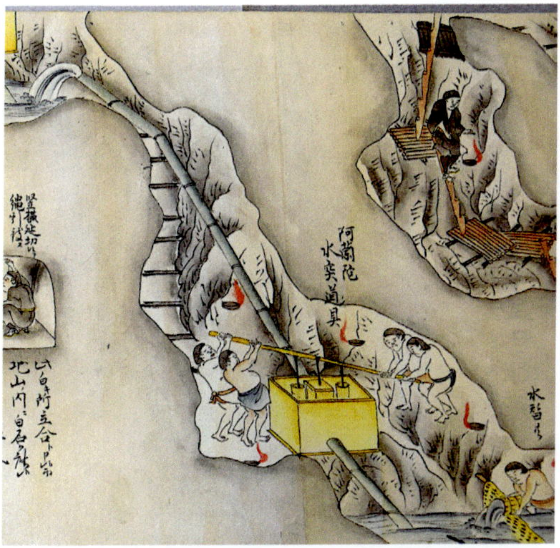

Fig. 9: Dutch style *brand spuijt*
This kind of device was brought to Sado in 1782 and used for underground drainage for a short period.
Source: Picture scroll *Kinginzan shikioka kasegikata zu*, inv. no. 12.24:35. Courtesy Museum Am Rothenbaum Kulturen und Künste der Welt, Hamburg; (Photo Regine Mathias).

35 Fumoto, *Sado kinginzan shiwa*, pp. 138–139.
36 Nagase-Reimer, 'Water Drainage in the Mines in Tokugawa Japan', pp. 35–36.

From 1777, the government in Edo had begun to round up and arrest *mushukunin* (lit. homeless people), who had often fled famines in rural areas and lived in the big cities unregistered, and therefore illegally in the view of the government. Many of them were deported to the gold mines of Sado, where they had to work as water drainage labourers.[37] This kind of conscripted labour helped save labour costs and was ultimately much cheaper to use than other means.

Shimin fuzoku (Manners and Customs of the Four Classes) notes that, after examining various devices such as *supontoi*, *suishorin*, and *brand spuijt*, the use of a bucket and a pulley block was the best for deep pits because of its infrequent failures and low cost. The book suggested that the *suishorin* would better serve farmers, who were not used to working with a pulley block.[38] As early as the 1670s, it was decreed that out-of-use *suishorin* should be given to farmers, who were to use them in agriculture.[39]

Drainage is vital for the mining industry. It is interesting to note that, while new equipment for drainage was introduced, which enhanced the general knowledge about technical solutions for this problem, ultimately, the mines returned to simple methods, which were less efficient, but cheaper and more reliable. It is likely that the preference for labour-intensive production was only in part due to the availability of cheap labour or the shortcomings of various devices. Another factor behind the preference for labour-intensive production could have been the fact that mines often made profits from selling food (especially rice) above purchase price to workers and their families, who lived in the mine compound. A larger population in the mines therefore increased the operator's income.[40]

In Europe, the need for drainage and the lack of cheap labour led to the use of water wheels underground. These were first powered by water and later by steam, a development which was one of the triggers of the Industrial Revolution. A range of devices and methods were used in Japanese mines at different times. These devices and methods seem to have been chosen in response to technical, managerial, political and social circumstances, and were considered to be adequate at the time they were used.

---

37 Isobe Kinzō, *Mushukunin – Sado kinzan hishi* (Unregistered people: The secret history of Sado), Tōkyō: Jinbutsu ōraisha, 1964.

38 Cited in Fumoto, *Sado kinginzan shiwa*, p. 244.

39 Fumoto, *Sado kinginzan shiwa*, p. 171.

40 Nagase-Reimer, 'Water Drainage in the Mines in Tokugawa Japan', pp. 38–39.

4. DISSEMINATION OF KNOWLEDGE

Mining knowledge was disseminated both by people and through writing. One must distinguish between horizontal transfer of knowledge, i.e., from outside to Japan and within Japan between the mines on the one hand, and the vertical transfer, i.e. within an employment relationship (training) on the other.

As we have seen, new knowledge flowed to Japan from China and Korea, but also from the West, during the sixteenth and seventeenth centuries. Sources contain many references to people who passed on this knowledge within Japan. For example, it is reported

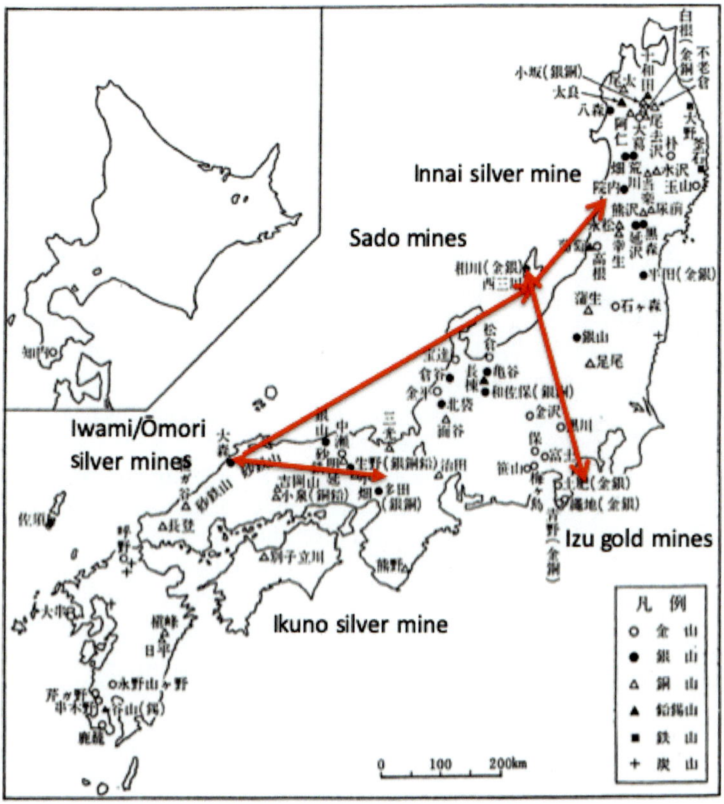

Fig. 10: Spread of mining knowledge
Exchange of persons and knowledge between major silver mines from late-sixteenth to early-seventeenth centuries, during the administration of mine director Ōkubo Nagayasu (1545–1613).

in the *Ikuno ginzan kyūki* (Old Reports on Ikuno Silver Mine) that it was mining 'experts' from Iwami, among them miners and smelters, who helped to develop the mine after its opening in 1542.[41]

A detailed study of the exchange of knowledge between mines reveals that the transfer of knowledge was also actively encouraged through the exchange of mining officials.[42] This was especially true for the period around 1600, and at mines, such as Iwami, Sado, and Ikuno, which were controlled by the Shōgun Tokugawa Ieyasu. Mines owned by feudal domains (*han*), such as Innai in the Tōhoku area, were also involved in this exchange of knowledge (Fig. 10).

A central figure in this exchange of knowledge around 1600 seems to have been Ōkubo Nagayasu (1545–1613), who was appointed magistrate for silver mining (*ginzan bugyō)* in 1601, and who administered mines in Iwami, Sado and Izu for many years on behalf of the Tokugawa government. When Ōkubo took charge of Sado, he brought experienced miners (*yamashi*) from other regions and used their knowledge to improve their output. Moreover, at his instigation, officials who had been involved with mining in the administration of the mines were sent back and forth several times to transfer knowledge and practices from Iwami to Sado and advance the mining there. Other officials and workers came from Kai, the home province of Ōkubo, and took their knowledge of surveying and rock processing to the development work in Sado.

When the gold and silver mines on the Izu Peninsula came under Tokugawa control after 1600, Ōkubo sent skilled officials from Sado to Izu to introduce new skills in dressing and smelting gold, which had become technically mature in Sado. Later on, experts from Iwami and Sado also visited the silver mine in Innai, which was managed by Akita domain (present-day Akita prefecture) in the Tōhoku area, to share their knowledge with fellow experts there. This exchange between the mines could be regarded as a kind of 'best practice' process, whereby miners and mines with the most advanced knowledge of certain processes were selected to advance mining in other places.

---

41 Cited in Nakano, *Ginzan shakai no kaimei*, p. ii.

42 Tsuchiya Hiroko, 'Tokugawa seiken no naritate to kinginzan. Kōzan ni okeru idō to kōryū kara' (The formation of the Tokugawa government and gold and silver mines. Mobility and exchange between mines), in *Hirosaki daigaku kokushi kenkyū* no. 113, 2002, pp. 24–44. The following passages are a summary of this study.

A much later example of such an exchange dates from the beginning of the nineteenth century. In 1813, a master smelter from Sado, who was well versed in the method of separating gold and silver, went to the Innai silver mine in northeast Japan and introduced the method there. This technique, which had already been applied in Sado since 1645, used sulphur as a flux prior to separating the silver sulphide from the gold. In 1838, 15 years after the smelter's death, his son also moved to Innai from Sado, taking new knowledge and an improved cementation technique to the mine. Subsequently, the amount of extracted silver is said to have increased significantly.[43]

These examples show that dissemination and knowledge transfer between certain mines by individuals played an important role in the development of ore mining during the Edo period. Even though ordinary miners were, in principle, not allowed to leave mines without permission, a degree of mobility among skilled miners and smelters was encouraged, so as to spread their knowledge as widely as possible.

Knowledge transfer took place not only between mines, but also *within* mines. Throughout the Edo period, knowledge was passed on within mines from experienced miners to younger workers. Young miners often started as haulers at the age of thirteen or fourteen. After four or five years of gaining experience, they became fully-fledged miners.[44]

The *kanabori,* a kind of 'foreman', had knowledge not only of extraction of the ore, but also of running of the ore veins, the topography of the rocks, as well as drainage and ventilation. The *kanabori* was responsible for workers in his group and acted as their *tsuchi-oya*, or 'hammer father'. The knowledge he passed on to young workers on the job was mainly based on his personal experience and information obtained from other miners. Another means of passing on knowledge was through stories about successful miners, but such stories usually focussed more on moral behaviour rather than on actual mining practice.

This type of on-the-job training and learning by doing was no different from training in other trades during the Edo period.

---

43 Ogi Shinichirō, *Kinsei kōzan wo sasaeta hitobito* (The people who supported mining in early modern Japan), Tōkyō: Yamakawa shuppan, 2012, pp. 66–67.

44 Regine Mathias, 'Gold und Silber fuer den Shogun. Japanische Bergleute: zum Profil einer sozialen Randgruppe in der Edo-Zeit', in Stephan Koehn and Chantal Weber (eds), *Outcasts in Japans Vormoderne. Mechanismen der Segregation in der Edo-Zeit*, Wiesbaden: Harrassowitz, 2019, pp. 177–200, p. 185.

The knowledge contained a large element of tacit knowledge, and it is therefore difficult to say whether a certain systematization or standardization had already developed during the Edo period.

While the oral, personal transfer of knowledge certainly dominated the transfer of mining knowledge in the Edo period, there was also a considerable written and pictorial tradition.

As early as in the seventeenth century, mining knowledge was compiled in manuscripts and printed books. Their number grew over time, peaking in the early nineteenth century. Their authors were mainly local officials, who were responsible for the mines and who had often collected their knowledge directly from miners. Books and manuscripts passed on knowledge through written descriptions of work processes, and pictures of equipment, such as furnaces and tools. Therefore, these books contained practical knowledge that was often directly related to specific mines and their circumstances, although they also included more universal explanations. Theoretical sections in the writings were still mainly based on Chinese ideas of *yin* and *yang*, the five elements, and mostly speculative knowledge.[45] Another source of knowledge were numerous picture scrolls depicting in detail the mining process, the various stages of processing the ore, the making of gold coins (*koban*) and the life in the mines.[46]

The relatively rich literature and pictorial material on mining in Japan contrast with China, for example, where only a few books on mining appeared prior to the early-nineteenth century. The main works in China were the *Tien Kung Khai Wu* (Exploitation of the works of nature) in 1637 (published as *Tenkō kaibutsu* in Japan in 1771) and the *Tien-nan Khuang-Chhang Thu Lueh* (Illustrated account of the mines and smelters of Yunnan) published in 1843–45.[47]

45 Several of these texts on mining were reprinted in the series *Nihon kagaku koten zensho*, edited by Saigusa Hiroto, vols. 9, 10, 11. Tōkyō, Ōsaka: Asahi shinbunsha 1942–44; a second reprint appeared in 1978. Others are included in the first two volumes of the series Nihon kōgyō shiryō shū kankō iinkai (ed.), *Nihon kōgyō shiryō shū, dai ikki and dai niki kinsei-hen* (Collected materials on Japanese mining history, vols. 1 and 2, early modern period), Tōkyō: Hakua shobō, 1981.

46 See e.g. Regine Mathias, 'Picture Scrolls as Historical Source on Japanese Mining', in Nanny Kim and Keiko Nagase-Reimer (eds), *Mining, Monies, and Culture in Early Modern Societies. East Asian and Global Perspectives*, Leiden, Boston: Brill, 2013, pp. 291–311.

47 Peter J. Golas, Joseph Needham, *Science and Civilisation in China*, pp. 23, 32–33.

Gold, silver and copper mining reached a high technical level throughout Japan in the early Edo period. However, with the short-lived exception of copper mining, mining production shrank dramatically and then stagnated at a very low level from the middle of the seventeenth century. Deposits were depleted or could no longer be reached with available technical means. The primary challenge was the lack of adequate methods for drainage. However, another obstacle was a lack of new knowledge that would have helped develop new technologies for exploiting remaining ore deposits. It was only the arrival of Western mining engineers after the opening of the country in 1854 that led to a short-lived revival in Japan's mining industry.

5. CONCLUSION

Until the beginning of the sixteenth century, the mining of precious metals in Japan was largely based on knowledge that was imported from China and Korea in the seventh and eighth centuries. This knowledge was supplemented in Japan over the following centuries. In the sixteenth century, knowledge in various areas related to mining rapidly developed, starting with the introduction of a new smelting process that turned Iwami mine into a centre of knowledge production. At that point, the country was open to knowledge from abroad and many new mines were opened based on the new knowledge. At this time we can also observe the formation of an expanding group of mining experts, comprising mining entrepreneurs and officials, who became the main agents in the exchange of knowledge through oral and, from the late-seventeenth century, through written and pictorial means.

Various political authorities actively supported the acquisition of mining knowledge. However, with the seclusion policy, existing knowledge was systematized and enriched by drawing directly on local reports and circumstances, with little ground-breaking new knowledge generated. Where there were technical improvements, these focused on relatively minor details. What was missing were schools, such as existed in Europe, where mining knowledge was not only taught in a more systematic form, but also substantiated by scientific research. Therefore, one crucial reason for the lack of new technological knowledge was the limited progress in scientific knowledge, evolving from and shaping the relationship between humans and nature.

REFERENCES

Azuma Kiyoshi, 'Tsushima no kuni kōginki to sono seirenhō' (The 'Tsushima kuni kōginki' and its smelting method), in: *Nihon kōgyō kaishi* 91 (1051), 1975–9, pp. 607–609.

Fumoto Saburō, *Sado kinginzan shiwa* (Historical Stories of the Sado Gold and Silver Mine), Tōkyō: Mitusbishi kinzoku kōgyō, 1973. (An abridged English version under the title *Historical Stories of the Sado Gold and Silver Mine* was published by Mitsubishi Materials Corporation in 2003).

Golas, Peter J., Joseph Needham, *Science and Civilisation in China,* vol. 5, *Chemistry and Chemical Technology,* part XIII: *Mining,* Cambridge: Cambridge University Press, 1999.

Lewin, Bruno (ed.), *Kodō zuroku. Illustrierte Abhandlung über die Verhüttung des Kupfer 1801,* (original by Masuda Tsuna, 1801). Facsimile reprint Bochum: Deutsches Bergbau-Museum, 1984.

Imai Noriko, 'Copper in Edo-Period Japan', in Keiko Nagase-Reimer (ed.), *Copper in the Early Sino-Japanese Trade*, Leiden, Boston: Brill, 2016, pp. 10–31.

Isobe Kinzō, *Mushukunin – Sado kinzan hishi* (Homeless people: The secret history of Sado), Tōkyō: Jinbutsu ōraisha, 1964.

Kanzaki Masaru, *Yakin kōkogaku gairon* (Outline of archaeometallurgy), Tōkyō: Yūsankaku, 2006.

Kobata Atsushi, *Nihon kōzan shi no kenkyū* (Studies in Japanese mining history) Tōkyō: Iwanami shoten, 1968.

Kurosawa Motoshige, *Kōzan shihō yōroku* (A digest of cherished treasures on mining), in Saigusa Hiroto (ed.), *Nihon kagaku koten zensho*, vol. 10, part 3, *Sangyō gijutsu hen. Saikō, yakin 2*, Tōkyō, Ōsaka: Asahi shinbunsha, pp. 1–75.

Mathias, Regine, "Gold und Silber fuer den Shogun. Japanische Bergleute: zum Profil einer sozialen Randgruppe in der Edo-Zeit", in Stephan Köhn and Chantal Weber (eds), *Outcasts in Japans Vormoderne. Mechanismen der Segregation in der Edo-Zeit*, Wiesbaden: Harrassowitz, 2019, pp. 177–200.

Murakami Ryū, *Kin gin dō no Nihonshi* (The Japanese History of Gold, Silver and Copper), Tōkyō: Iwanami shoten, 2007.

Nagase-Reimer, Keiko, 'Water drainage in the mines in Tokugawa Japan: Technological improvements and economic limitations', in Nanny Kim and Keiko Nagase-Reimer (eds.) *Mining, Monies, and Culture in Early Modern Societies. East Asian and Global Perspectives*, Leiden, Boston: Brill, 2013, pp. 26–42.

Nakanishi Tetsuya, Izawa Eiji, 'Evolution of Silver-smelting Technology on Japan in the Middle of the Sixteenth Century', in *ISIJ International*, vol. 54 (2014), no. 5, pp. 1093–1097.

Nakano Yoshifumi, *Ginzan shakai no kaimei. kinsei Iwami ginzan no keiei to shakai* (Shedding new light upon mining society. Society and management of the early modern Iwami silver mine), Ōsaka: Seibundō, 2009.

Nishio Keijirō, *Nihon kōgyō shiyō* (The essentials of Japanese mining history), n.p.: Jūichi-gumi shuppanbu, 1943.

Ōsaka Museum of History (ed.), *Tokubetsu ten: Yomigaeru akagane. Nanbanfuki to Sumitomo dōfukisho* (Special exhibition: Rethinking copper. The *nanbanfuki*-method and the Sumitomo copper refinery), Ōsaka Museum of History, 2003.

Ogi Shinichirō, *Kinsei kōzan wo sasaeta hitobito* (The people who supported mining in early modern Japan), Tōkyō: Yamakawa shuppan, 2012.

Smith, Cyril Stanley, 'Introduction', in Smith, Cyril Stanley (ed.) *Kodō zuroku, Illustrated Book on the Smelting of Copper* by Masuda Tsuna (facsimile edition with English translation by Zenryu Shirakawa), Norwalk, Connecticut: Burndy Library, 1983, pp. 9–25.

Suzuki Kazuyoshi and Tanabe Yoshikazu, 'Edo shoki no hōi oyobi kakudo no gainen kara mita sokuryōjutsu no keisei nit suite no ichi kōsatsu' (The concepts of 'angle' and 'direction' in schools of surveying method in early Edo period) in *Bulletin of the Museum of Natural Science,* Ser. E, 32, 2009, pp. 41–49. (http://www.kahaku.go.jp/research/publication/sci_engineer/v32.html, accessed 10.09.2018)

Tanaka Keiichi, 'Edo shoki amarugamu-hō no dōnyū to Ieyasu no bōeki seisaku' (The introduction of the amalgamation method in early Edo period and Ieyasu's trade policy), in *Nihon rekishi* 501 (1990.2), pp. 71–84.

Tsuchiya Hiroko, 'Tokugawa seiken no naritate to kinginzan. Kōzan ni okeru idō to kōryū kara' (The formation of the Tokugawa government and gold and silver mines. Mobility and exchange between mines), in *Hirosaki daigaku kokushi kenkyū* 2002–10, 113, 2002, pp. 24–44.

Walter, Lutz (ed.), Japan mit den Augen des Westens gesehen. Gedruckte europaeische Landkarten vom fruehen 16. bis zum 19. Jahrhundert. Munich, New York: Prestel 1993.

Yoshioka Toya, *Iwami ginzan wo yomu. Kozu, emaki, kyūki, sekishūgin* (Reading Iwami silver mine. Old maps, picture scrolls, ancient reports, silver from Sekishū), Izumo: Hōkōsha, 2017.

5

Tanaka Hisashige and His Myriad Year Clock: Its Technological Characteristics and Historical Background

HASHIMOTO Takehiko

1. INTRODUCTION

TANAKA HISASHIGE IS well known today for his various ingenious inventions and for his construction of useful or entertaining machines, particularly his contrived automata dolls (*karakuri ningyō*). He is also known for his later involvement in the modernizing attempts of the Saga domain (present-day Saga prefecture) and as one of the founders of the company that is now Toshiba. He should also be remembered as the craftsman who designed and made the intricate machine called the 'Myriad Year Clock' (Fig. 1), or *Mannen dokei*, a masterpiece of Japanese craftsmanship, before the systematic introduction of Western technologies.

Fig. 1: The Myriad Year Clock (Photo: Toshiba Company).

In this paper, I explain the mechanical structure and technological characteristics of this mechanical masterpiece based on my experience and from information acquired during a project to disassemble and restore the clock.[1] I first provide a short biography of Hisashige, as well as a brief explanation of the project of disassembling the clock. I then explain the basic structure of the Myriad Year Clock together with technological features observed by the engineers engaged in the disassembling and restoration project.[2]

2. A BIOGRAPHY OF TANAKA HISASHIGE

Tanaka Hisashige, originally named Tanaka Giemon, was born in 1799 as the eldest son of a tortoiseshell craftsman.[3] Early in his childhood, he showed mechanical ingenuity in making devices, some of which were useful to his neighbours. At the age of eighteen, his father passed away. Although he was expected to succeed his father's position, he conceded it to his younger brother,

---

1 The activity and consequent findings of this disassembling and restoring project will be narrated in Section 3 and later sections. The reports concerning this project are contained in Hashimoto Takehiko (ed.), 'Tokei no gijutsuteki tokuchō to shakaiteki igi ni kansuru rekishiteki kenkyū (Historical researches on technological characteristics and social significance of the Myriad Year Clock)', an unpublished report of the Grant-in-Aid for Scientific Research of the JSPS (Japan Society for the Promotion of Science) 2006; the report is available at the National Diet Library, Tōkyō; for details see https://ndlonline.ndl.go.jp/#!/detail/R300000001-I000008447651-00. I have discussed some of the findings of this project in Hashimoto Takehiko, 'Mechanization of Time and Calendar: Tanaka Hisashige's Myriad Year Clock and Cosmological Model,' *UTCP Bulletin*, 2006, vol. 6, pp. 47–55.

2 I will mostly rely on the observations of the engineers of Seiko (Seiko Holdings Corporation), Tōkyō, engaged in the project. Toshiba engineers engaged in the project also contributed findings on the technical characteristics of the Myriad Year Clock. One of these Toshiba engineers completed a Ph.D. dissertation based on their analysis of the clock. See Kinoshita Yasuhiro, 'Mannen Dokei no bunkai chōsa kara mita Bakumatsu seiyō gijutsu juyō ni kansuru kenkyū: Mannen Dokei no gijutsuteki genkai to Tanaka Hisashige no gijutsusha e no tenkan (A research on the acceptance of Western technologies at the end of the Edo period seen from the disassembling investigation of the Myriad Year Clock: the technological limitation of the Myriad Year Clock and the transformation of Tanaka Hisashige into an engineer),' Ph.D. dissertation, Tōkyō Institute of Technology, 2015. The author's name was Yokota Yasuhiro at the time of the project.

3 There are several biographies of Tanaka Hisashige. See Tanaka Ōmi Ō Kenshō kai (eds), *Tanaka Ōmi Daijō*, Tōkyō. 1931, and Imatsu (Imazu) Kenji, *Karakuri Giemon: Tōshiba sōritsusha Tanaka Hisashige to sono jidai* (The ingenious Giemon: the life and times of Tanaka Hisashige, the founder of Tōshiba), Tōkyō: Daiyamondo-sha, 1992.

deciding instead to become an inventor. Thereafter, he devised numerous entertaining automata, including a teacup-carrying doll, an arrow-shooting doll, and a doll that wrote Chinese characters. He displayed them at shrines or other public spaces, a common practice in various cities in Japan since the late eighteenth century.

Thereafter, Hisashige travelled widely around western Japan, eventually settling in Ōsaka and then Kyōto where he opened a shop to make various machines. At the Kyōto shop, he also started to repair damaged clocks, and this business turned out to be a success. In addition, he invented a new type of oil lamp with a glass shade, as well as a pressurized pump to supply oil from below. The two contrivances made the lamp's illumination brighter and more stable than traditional candle lamps. This lamp, called *mujintō* (eternal lamp) became very popular and was sold widely. Tanaka Hisashige manufactured several different types of *mujintō* and succeeded in making a tidy profit from this business as well.

With the profits from clock reparation and *mujintō* sales, Tanaka Hisashige undertook the construction of the Myriad Year Clock. a relatively large, complex mechanical system with precisely-made components. The clock comprises six faces and a solar and a lunar working model on the top. The six faces on the hexagonal main body included Japanese and Western clocks, as well as a calendar and a lunar appearance display. This became Tanaka Hisashige's lifework and was completed in 1851. The Myriad Year Clock is made of more than one thousand mechanical parts, and contains two clocks and self-moving astronomical model.

3. THE DISASSEMBLING PROJECT

In 2004, a large collaborative project was launched to disassemble this clock-work system, restore its operating condition, and construct replicas. The project was sponsored by Toshiba Co. Ltd. as the object of study was the masterpiece of one of their founders. Technical work on the project was carried out mainly by professional horological engineers of Seiko Holdings Corporation using their facilities at the Makuhari unit of the Seiko Instruments Company. Most were retired Seiko engineers, who were experts in making and fixing mechanical watches. The project was conducted as a part of a larger project entitled 'Inventions in the Edo Period', which was sponsored by the Japan Society for the Promotion of Science (JSPS), and led by Suzuki Kazuyoshi at the National Science Museum.[4] As a part of this large JSPS project, I organized a smaller collaborative project on the history of clocks and participated in the disassembling and replication as an observer and a recorder of the engineering project.

The replica made through this project was to be displayed at the World Exhibition, held in Nagoya in 2005. Because of the deadline set by this plan, the group of the collaborative project and the engineers were forced to omit some of technical work required to fully examine the mechanical construction of the Myriad Year Clock.[5] After completion of the technical work, I interviewed the engineers involved about their observations and findings on the mechanical characteristics of the Myriad Year Clock.[6] Based on this information, I compiled a report that also included explanations on the internal structure of the clock.[7]

4. THE STRUCTURE OF THE MYRIAD YEAR CLOCK

The Myriad Year Clock is divided into two main parts—the upper clock-work machine, and the lower hexagonal case containing four powerful coil springs. The upper part comprises the astronomical model on the top (Fig. 2), and the hexagonal box housing two clocks and calendrical displays. The hexagonal box has six faces, each showing different time-related information. The European-made watch, which was identified in the project as the first of the six faces, serves as the fundamental time-keeping unit in the entire clock-work system. The other five faces comprise the following: a Japanese clock of the *warikoma dokei* type (Fig. 3); a manually rotating indicator which identifies the twenty-four seasonal divisions

4 The project attracted the interest of the NHK (Japan Broadcasting Corporation), and its staff also participated in it. They recorded two video project activities throughout the year and made a special TV program titled 'Mannen dokei: Edo-jidai no unda kyōi no kikai dokei' (The Myriad Year Clock: a miraculous mechanical clock generated by a genius in the Edo period), which was broadcast in 2005.

5 Tsuchiya Hideo, 'Tanaka Hisashige 'Mannen Dokei' chōsa hōkoku' (A Report on the Investigation of 'the Myriad Year Clock' of Hisashige Tanaka), *Micromechatronics*, vol. 50, no. 194, 2008, pp. 58–68.

6 The engineers I interviewed were Tsuchiya Hideo, Kessoku Yoshinosuke, Oishi Masaaki, and Muraji Shigenori. The former three were retired Seiko engineers, and the latter was an active Seiko employee.

7 Hashimoto Takehiko, 'Mannen dokei kikōbu no bunkai chōsa hōkoku' (A report of disassembling investigation of the Myriad Year Clock), Chapter 5 of the JSPS report, pp. 51–112.

Fig. 2: The astronomical model on the top of the Myriad Year Clock (Photo by the author).

Fig. 3: The Japanese *warikoma* clock (Photo: Toshiba Company).

and their starting dates; an indicator of the seven days of the week together with an indicator of Japanese time; the indicator of the ten calendrical and twelve zodiac signs (the combination of which indicate signs of the sexagenary cycle); and the indicator of the phase and the day of the month of the lunar calendar.[8] Of these

8 Tsuchiya provides succinct explanations and engineering comments on the mechanisms on and behind these six faces; see Tsuchiya Hideo, 'Tanaka Hisashige

six faces, only the third face is manually adjusted. The indicators on all of the other faces are operated by more or less complicated mechanisms. In summary, the Myriad Year Clock consists of the astronomical model at the top, six blocks at the middle box, and the driving block at the base.

Fig. 4: The internal view from the top (Photo by the author).

Fig. 5: The internal view from the side (Photo by the author).

'Mannen dokei' chōsa hōkoku' (A report on the investigation of 'the Myriad Year Clock' of Hisashige Tanaka), *Micromechatronics*, vol. 50, no. 194, 2008, pp. 58–68, esp. pp. 60–64.

When the astronomical model is removed from the top of the hexagonal box, it reveals various gear wheels encased within (Fig. 4). Another view with the European watch removed shows that the gears are arranged in three levels (Fig. 5). It shows thick gears at the lower level and thinner and finer gears in the middle and upper levels. It also allows a glimpse of the intricate mechanism behind the *warikoma*-type clock on the left.

In the project, the Seiko engineers prepared precise drawings for all the components of the clock system so that they could construct a replica. They also prepared a diagram of the entire mechanism (Fig. 6), containing powerful spring motors

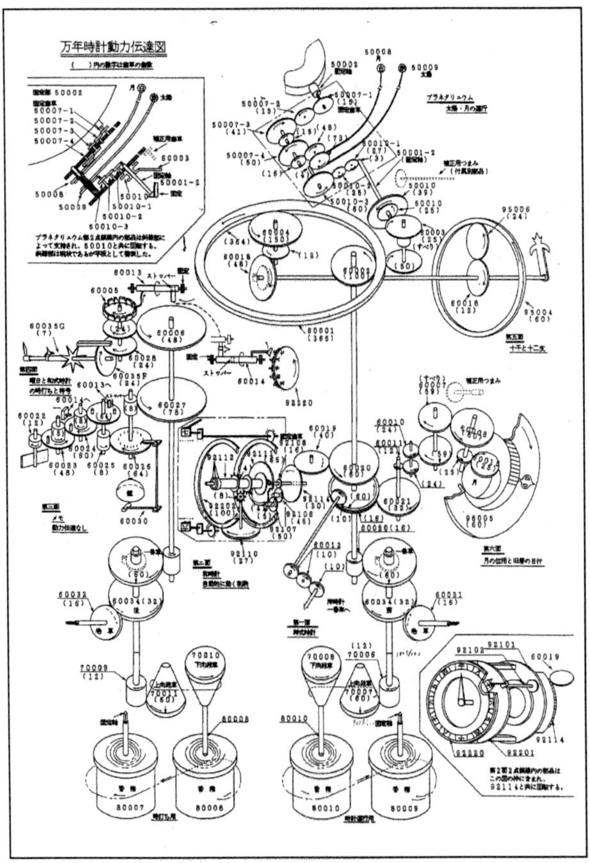

Fig. 6: The scheme of the entire mechanism (Diagram drawn by the group of the Seiko engineers)

(Fig. 7a) at the bottom, and the astronomical clock at the top. In the middle, the diagram shows the combination of gears leading to the side faces of the hexagonal box. Notice that the motive sources are basically separated into two independent parts represented by the left and right vertical axes in the diagram. The right vertical axis, driven by the four spring boxes (Fig. 7b) and the two fusées (Fig. 7c), is connected to all time-keeping devices and their mechanisms, including the astronomical model at the top. The left vertical axis, on the other hand, moves the mechanism that rings

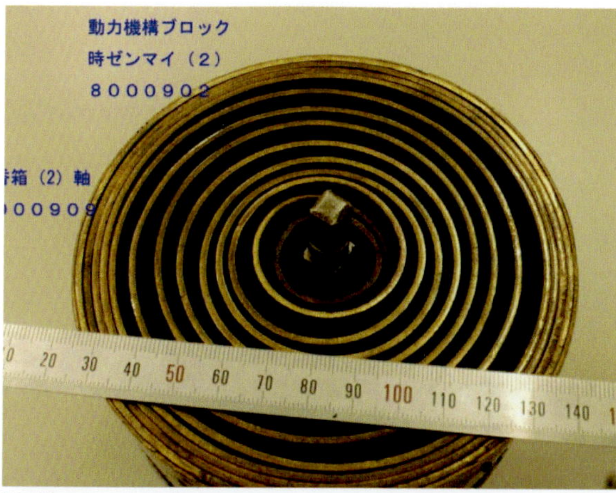

Fig. 7a

Fig. 7b

Fig. 7c: The powerful spring as a driving force. The source of driving force was located in the hexagonal case at the bottom. The four powerful springs made by brass plates (7a) are housed in four cylindrical boxes (7b), forming two pairs, each of which drives the entire clock system or the system to ring a bell at specific times a day. The two axes separately driven by these two pairs are connected to the fusée mechanisms at the upper level (7c), through which the driving forces become more uniform over time (Photo by Yamakoshi).

the bell. The timing of the bell ring is controlled by the time-keeping device.

I now direct attention to the central part of the time-keeping mechanism along the right axis (Fig. 8). It shows two separate yet connected vertical axes driven by the spring hidden below. The first wheel of the upper axis connects to the European watch on the first face. The mechanism inside the European watch is omitted in the diagram. This imported clock is a mechanism given to Hisashige so we were less concerned about its mechanism. The gear directly connected with the European watch rotates once every eight hours. The upper axis is controlled so that it rotates once every forty-eight hours. The upper axis is connected to the Japanese *warikoma* clock through the gears numbered 60020, 60019, and 92114, which comprise 60, 40, and 30 gears, respectively. Through these gears, the period of 48 hours is accelerated to 24 hours so that its indicator rotates once per day.

On the right side, the same gear wheel (60020) connects the gear 60021 to the block showing the lunar phase (Fig. 9).

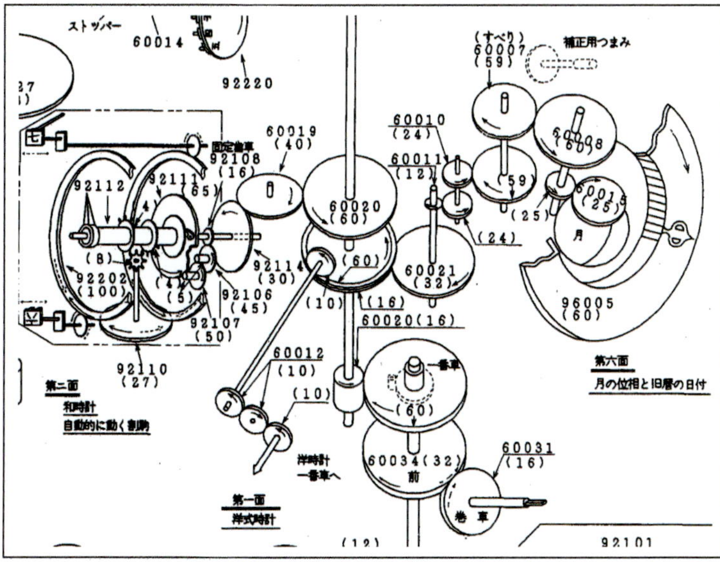

Fig. 8: The central part of the scheme of the internal mechanism. A magnified view of Fig. 6.

Fig. 9: The ball representing the lunar eclipse (Photo by the author).

The transmission involves ten gear wheels between the motive shaft and the ball for the lunar phase. The simple calculation of the ratios of the numbers of adjacent gears shows that the rotational speed of the motive shaft is geared down to one rotation every 20 days at the ball of the lunar phase, as follows:

$$2 \text{ days} \times \frac{32}{16} \times \frac{24}{12} \times \frac{59}{24} \times \frac{60}{59} \times \frac{25}{25} = 20 \text{ days.}$$

This, however, is not right. This fact was noted separately and almost simultaneously by Seiko and Toshiba engineers when they were analysing the internal mechanism of this clock.[9] It is uncertain whether this mistake was made by Tanaka Hisashige himself or later by craftsmen who attempted to analyse and fix the clock. The use of a gear wheel with 59 teeth suggests that Hisashige tried to use the 29.5 days as a cycle of one lunar month. (To attain this number, the two-gear component 60007 should have 40 teeth instead of 59 for its upper gear connecting with the gear 60008.) Even if it was Tanaka Hisashige's decision, he likely did not intend such a design flaw.

The connection between the upper axis and the astronomical model is complicated. The connection is intermediated by the large gear wheel 60001, displayed in the view from the top (Fig. 4). The upper axis from the motive power reaches the gear 60002 and, intermediated by this large gear wheel, connects the two gears—gear 60004 inside, and 60003 outside. Gear 60004 leads to the block of the sexagenary cycle, and the gear 60003 leads to the astronomical model above. Gear 60002 has 100 teeth and gear 60003 has 50; therefore, the rotating speed of the motive axis is accelerated by these gear connections from a 48-hour period to about a 24-hour period. This speed is transmitted to the double disk that rotates once per day around the solar and lunar balls. The intervention of the large gear wheel 60001, however, brings a complication to this speed change. It has 365 teeth on the outer rim, but 364 teeth, just one tooth fewer, on the inner rim.[10] The exact period of the gear 60003 is slightly shorter than 24 hours at about 23 hours and 56 minutes. As I will explain later, this is because the sun stick and ball rotate one turn every 365 days by the internal mechanism, and the slight reduction of the rotational speed of the astronomical discs allows the sun to rotate with an exact 24-hour period.

---

9 Hashimoto Takehiko, 'Mannen dokei kikōbu no bunkai chōsa hōkoku' (A report of disassembling investigation of the Myriad Year Clock), Chapter 5 of the JSPS report, esp. pp. 108–111.

10 Tsuchiya Hideo mentions this large gear wheel and the necessity of using such a large number of cogs to gear down the required rotational speeds.

5. THE MECHANISM OF THE JAPANESE *WARIKOMA* CLOCK

The most intricate part of the Myriad Year Clock is the Japanese *warikoma* clock on the second face. To explain its mechanism, it is first necessary to explain the time system in the Edo period.

The time system in the Edo period was similar to ancient and medieval time systems in the West, whereby units of time changed length from daytime tonight, as well as from season to season. Such a variable time counting system is called a 'seasonal time system', in contrast to a 'clock-time system' (also called 'fixed-time system'), the system that we use today. The day was divided into twelve units of time (*shinkoku*), with six units for daytime and six for the night. The twelve units of time were named according to the twelve zodiac signs of the traditional Chinese counting system, or they were labelled with the six numbers from 4 to 9, which were each used twice, before and after noon.[11] Midnight was known as the time of the rat or the 9th hour, dawn as the time of the rabbit or the 6th hour, noon as the time of the horse or the 9th hour, and dusk as the time of the rooster or the 6th hour.[12]

Figures 10a, 10b, and 10c show faces of this *warikoma* clock at three different seasonal days. The clock it shows is the original one but it has been detached from the box, disassembled and readjusted to indicate the proper seasonal time as well as seasonal division. The entire circle of the clock rotates once per day so that the *warikoma* indicator at the top, the character of the horse in this case, shows the current hour of the day based on the seasonal time system. The single short hand of this clock indicates the seasonal division; the three photos show the faces at the winter solstice (Fig. 10a), the spring

11 On the time system in the Edo Japan and various types of mechanical clocks indicating its seasonal time system, see Mody, N.H.N., *Japanese Clocks*, Rutland, VT and Tōkyō: Charles E. Tuttle, 1932; Hashimoto Manpei, *Nihon no jikoku seido* (Time System of Japan), Tōkyō: Hanawa shobō, 1994, esp. pp. 123ff.; Yamaguchi Ryūji, *Nihon no tokei* (Clocks in Japan), Tōkyō: Nihon hyōronsha, 1950; Yulia Frumer, *Making Time: Astronomical Time Measurement in Tokugawa Japan*, Chicago: University of Chicago Press, 2018.

12 Pre-modern China used the same time-counting names. However, unlike Edo-period Japan, China adopted the clock (or fixed) time system, and its official time-counting system designated by the 9th hour (or the time of rat or of horse) was from 11 o'clock to 1 o'clock, not from 12 to 2. In Edo-period Japan, the official time-counting system adopted a similar method, but people usually adopted a different practical method whereby noon and midnight were designated by the 9th hour, and dawn and dusk by the 6th hour. All mechanical clocks, including the one in the Myriad Year Clock, also indicated time according to this practical time-counting system.

Fig. 10a

Fig. 10b

Fig. 10c: The face of the Japanese clock and the seasonally changing positions of its *warikoma* plates. The positions at winter solstice (10a), spring equinox (10b), and the summer solstice (10c) (Photo by Yamakoshi).

equinox (Fig. 10b), and the summer solstice (Fig. 10c), respectively. At the summer solstice, the spacing of the *warikoma* indicators become conspicuously different between the daytime on the upper horse side and the night-time on the lower rat side. The lengths of unit time between the daytime and the night-time are less different at the equinox and the winter solstice. Some readers might wonder why the lengths of hours in the daytime and the night-time at the winter

Fig. 11: The intricate mechanism behind the face of the Japanese clock (Photo by the author).

Fig. 12: The face of an ordinary *warikoma* type of clock (Oda (ed.), 1986, p. 75).

solstice is not as great as in the summer solstice or why they were not symmetrical at the two solstices. This is because the Japanese seasonal time system defines the beginning and end of the daytime not at the sunrise and sunset, but at dawn and dusk, thereby making the average length of a day slightly longer than the length of a night.

Behind this front face of the clock, there is a complicated mechanism consisting of numerous small and large gear wheels (Fig. 11). An ordinary *warikoma* clock (Fig. 12) needs a manual adjustment of all twelve *warikoma* indicators at every seasonal division. It therefore has a regularly calibrated circular scale on the outer rim so that its user can adjust the *warikoma* indicators easily on the precisely calculated positions. The *warikoma* clock of the Myriad Year Clock, however, has a mechanism to automatically move the *warikoma* indicators to their correct positions continuously all year round.

The *warikoma* clock of the Myriad Year Clock has several layers. Between the circular plate holding ten miniscule gears and the front plate holding *warikoma* indicators, there is a crank mechanism comprising oscillating groove holders (Fig. 13a) and rotating cranks that connect to tiny gears (Fig. 13b). Behind the *warikoma* face, there are ten moving pieces with grooves perpendicular to the circular slit, five connecting with *warikoma* indicators for morning hours and the other five with *warikoma* indicators for afternoon hours. The back side of the plate of tiny gears also holds ten small cranks of differing lengths (Fig. 13c), five of which mesh

Fig. 13a

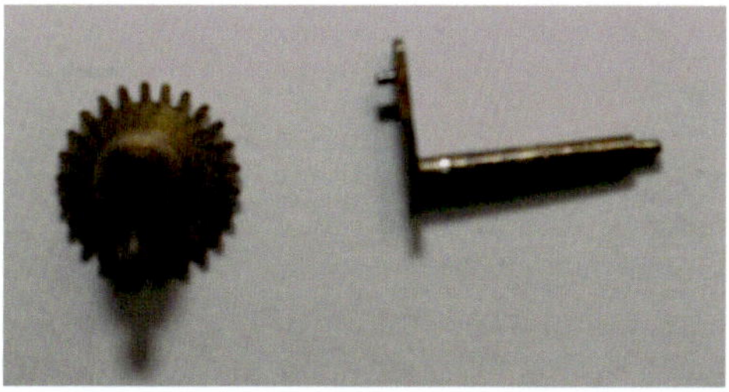

Fig. 13b

Fig. 13c: The cranks and grooves in the mechanism of the Japanese clock. Behind the face of the Japanese clock, there are 10 movable groove holders (13a), each of which moves *warikoma* plates on the front face. Their grooves mesh with the tiny pins of a crank whose longer rod pierces through a tiny gear and is fixed by a tiny screw bolt (13b). The cranks are connected on the internal plate and arranged along the circle of *warikoma* plates (13c) (Photo by Yamakoshi).

with the grooves for the morning *warikoma* and the other five with those for the afternoon *warikoma*.

The tiny gears make oscillatory rotation with the period of one year, thus the grooves and the *warikoma* indicators slowly move to complete one oscillation per year. Speaking more precisely, each tiny gear makes an oscillating rotation every half a year by Tanaka Hisashige's intricate and original mechanism that utilized gears with a peculiar shape. Along the short central axis in the middle of the *warikoma* clock, there are two incomplete gears that have teeth only on one side, one with three teeth on one side, and another with three teeth on the other side (Fig. 14a). These two gears vertically mesh with another strangely shaped gear (Fig. 14b). Because of its shape, it was first called '*dani haguruma*' (tick gear), but was later renamed '*mushi haguruma*' (insect-shaped gear). When the two semi-circularly toothed gears rotate one turn, the insect-shaped gear makes a half rotation in one direction and then goes on to make another half rotation in the opposite direction.

Fig. 14a

Fig. 14b: The two half-toothed gears of the Japanese clock (14a) and the insect shaped gear (14b) (Photo by the author).

Figure 15 shows a paper-made impromptu model of these three special gears. This was made by a Seiko engineer to display how the three gears mesh with each other to transform regular unidirectional rotation into an oscillatory rotation.[13] The two half-toothed gears are pierced and supported by the horizontal axis (made by a long chopstick) and rotated by the two hands of the engineer. The right half-toothed gear (positioned near the left hand of the engineer) is now about to start to mesh with the teeth of the insect-shaped gear — the teeth represented by the three left legs of the insect. While the four teeth of this right gear proceed to mesh with the legs of the insect, the head of the insect slowly turns from right to left; and when the two gears complete meshing, the left half-toothed gear (positioned near the right hand of the engineer) starts to mesh with the right legs of the insect. While the two gears proceed to mesh, the head of the insect now turns from left to right. Thus, while the chopstick axis makes one rotation, the insect turns its head from right to left and then from left to right in one oscillatory rotation.[14]

Fig. 15: The paper model to represent the engagement and working of the half-toothed gears and the insect-shaped gear (Photo by Yamakoshi).

13 Hashimoto Takehiko, 'Mannen dokei kikōbu no bunkai chōsa hōkoku' (A report of disassembling investigation of the Myriad Year Clock), chapter 5 of the JSPS report, esp. pp. 88–89.

14 The NHK TV program provides a video clip, which explains this mechanism clearly. Kinoshita also provides explanations with visual images, see Kinoshita op. cit, pp. 112–115.

With this intricate mechanism to transform rotation into annually oscillating rotation, Hisashige's Japanese clock was expected to move the *warikoma* plates slowly and automatically so that they indicated the time according to the seasonal time of Edo Japan. The measurement of cranks contained in this mechanism suggests that the indication of the seasonal time was fairly accurate, but not as accurate as other non-automatic clocks when adjusted by experts. The shortest cranks in symmetrical positions have the lengths of 2.03 and 2.19 mm, although they should be precisely equal in ideal construction.[15] The difference of the lengths indicates the level of accuracy in the actual construction of the Japanese clock of the Myriad Year Clock.

6. THE TECHNOLOGICAL FEATURES OF THE MYRIAD YEAR CLOCK

The central piece of the Myriad Year Clock, the compact automatic *warikoma* type of clock, was thus intricately designed and devised. Did it work as conceived in reality? The answer is most likely no. It might have worked for a short time after construction, but the clock is not functional today because of friction between the parts and for other mechanical reasons.[16]

Fig. 16a

[15] See Hashimoto, 'Mannen dokei kikōbu no bunkai chōsa hōkoku', esp. p. 90.

[16] See Hashimoto, 'Mannen dokei kikōbu no bunkai chōsa hōkoku', esp. pp. 55–56 and 64–70.

Fig. 16b: A miniscule gear (16a) and a miniscule screw (16b) used in the mechanism of the Japanese clock (Photo by Yamakoshi).

The Japanese clock contained millimetre-sized screws for the 10 miniscule gear wheels (Fig. 16a). The gear wheel has a cylindrical hole at its central axis, and the crank's cylindrical rod pierces through it. A millimetre-sized screw (Fig. 16b) is intended to fix the rod inside the gear wheel's hole. The Seiko engineers found that all these rods rotate loosely inside the holes since their screws are all damaged. In order to restore the original clock system, they decided to make new screws slightly larger than the original ones and to the holes as well. How was Tanaka Hisashige able to make such small screws? We do not know. A Seiko engineer, Tsuchiya Hideo, conjectures that Tanaka Hisashige probably used an iron-made screw and dug a screw hole with the iron screw. However, this raises an unanswered question: how did he make tiny iron-made screws?

Analysis of the Myriad Year Clock's disassembled parts has revealed several other examples of the technical skills and methods of making this clock-work machine.

Figure 17 shows the gear wheel numbered 60002 connecting the upper motive shaft and the large conspicuous gear wheel, which transmits the motive force to the astronomical model above it, as well as to the sexagenary indicator. Since the astronomical model houses many gear wheels together with solar and lunar components, the gear 60002 must transmit a large motive force, especially to the gear wheel 60003, which rotates the entire mechanism of the astronomical model. Probably for this reason, Tanaka Hisashige replaced a cylindrical shaft with a quadrangular one with a square section. It does withstand a large torque but has a few drawbacks as well. This makes it difficult for craftsmen to determine the true centre of the gear wheel or to cut a square

Fig. 17: The gear with square-shaped axis and traces dividing angles (Photo by Yamakoshi).

opening in such a way that its centre precisely corresponds with the centre of the circle of the whole gear wheel.[17]

This gear wheel is a rare sample that leaves the straight and circular lines marked with a knife edge. The marking clearly shows that the teeth of the gear were divided by hand, and that the cogs were all made by hand. The division of this gear is just one hundred angles. We do not know exactly how Tanaka Hisashige or other craftsmen in Tokugawa Japan divided a whole circle into different numbers of angles. Western engineers developed mechanical methods to divide the angle using the combination of screws and gears. There remains an instrument owned by an assistant of Tanaka Hisashige to make screws, but it was much simpler than Western ones. Hisashige might have used this kind of instrument as well to make screws or divide angles.

Setting aside the angle division, how were teeth cut to make the gear cogs? The cogs all appear to be quite precise. They are accurate, but how accurate? Another gear wheel with 48 teeth was observed under a microscope to measure the accuracy and deviation of the position and shape of several of its teeth. The diagram in Fig. 18 shows the extent of correspondence in shape and

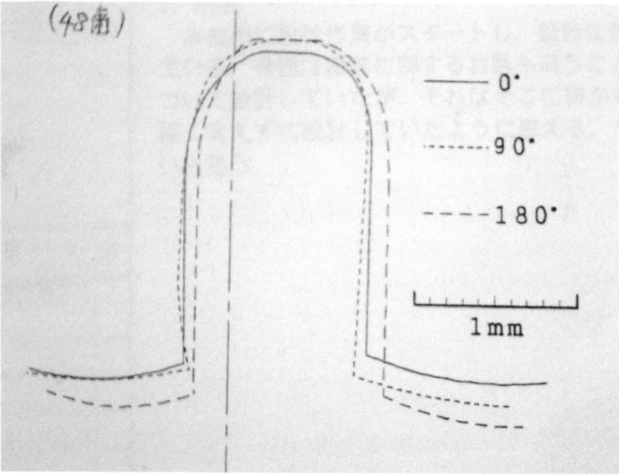

Fig. 18: The diagram to indicate the sub-millimeter discrepancy of the positions of the cogs of a wheel. The solid line indicates the cog at the specific position; the dotted line the cog at 90 degree rotated from the first cog; and the broken line the cog at 180 degree rotated from the first cog (Diagram by Muraji).

[17] See Hashimoto, 'Mannen dokei kikōbu no bunkai chōsa hōkoku', esp. pp. 72–73.

position of the three different teeth at the locations of a reference point, 90 degrees different from it, and 180 degrees different from it. From this diagram, the difference of the correspondence is estimated to be less than 0.2 mm. According to an engineer who participated in the disassembling project, the impressive large gear wheel located at the top of the clock system (the gear wheel 60001 in Fig. 4) also shows such sub-millimetre discrepancy under the microscope, proving that it too was hand-made.

Oishi Masaaki, the Seiko engineer in charge of the inspection of the design and mechanical construction of the Myriad Year Clock, has listed several other technological features of Tanaka Hisashige's clock system.[18] He points out that Tanaka Hisashige used cogs having only a few teeth, those having pins instead of cogs, and those that vertically engaged each other. The use of gears with a small number of teeth and its reason will be explained in the next section.

7. ASTRONOMICAL MODEL

The astronomical model at the top of the clock system represents the celestial hemisphere over the ground of Japan. It consists of the brass meridian circle, the red celestial equator, and the two rotating balls (red and silver). The red one represents the sun and the silver the moon. The two celestial balls rotate daily over the plane of a map of Japan, intricately changing their altitudes from season to season. Thus, the red ball, which represents the sun, shows up in the east along the red equator on the days of spring and autumn equinoxes, and its position changes to a higher altitude during summer and to a lower altitude during winter. The silver moon ball changes its altitude more frequently. The mechanism for their alteration in altitude lies under the hemisphere of the Earth.

When the the hemispheric Earth is taken away, it reveals a hemispheric concave bowl underneath (Fig. 19). On its right side, you will see a smaller double disc mechanism that rotates the silver and red balls daily and changes their altitudes monthly and seasonally. This sun-and-moon mechanism consists of two curved discs and an internal gearing mechanism housed between the two discs. The red ball and stick were moved by the gearing mechanism attached to the outer disc (Fig. 20), and the ball and stick annually rotate, not around the centre of the disc, but

---

18 See Hashimoto, 'Mannen dokei kikōbu no bunkai chōsa hōkoku', esp. pp. 71ff.

Fig. 19: The hemispheric bowl below the astronomical model (Photo by Izawa).

Fig. 20: The stick and red ball representing the sun (replica) (Photo by Izawa).

around the epicentre near the periphery of the disc. Therefore, the distance between the centre of the disc and the ball gradually changes throughout a year. The most distant position, as shown in Fig. 20, stands for the sun at the summer solstice, the two vertical positions at left and right for the two equinoxes, and the least distant position, opposite to the one at the summer solstice, for the winter solstice. (The distance at the two equinoxes is not the arithmetic mean between the two, although they should be so astronomically.)

The diagram (Fig. 21) shows the sun-and-moon's gearing mechanism with the number of cogs on each gear wheel. It specifically shows how the daily rotation is geared down to the annual rotation of the stick of the red solar ball. The diagram designates the numbers of cogs for each wheel, and the combination of these six gears reduces the rotating speed to 1 over 365, thus allowing the stick with the red ball to complete one rotation over 365 days. As the diagram shows, this gear reduction system contains those gears with only three cogs or four cogs. The replica does not have

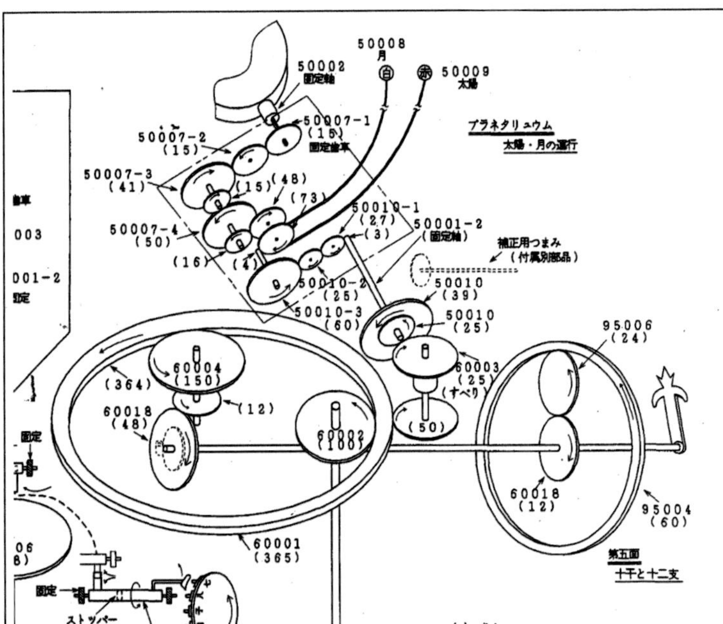

Fig. 21: The mechanical scheme of the astronomical model. This is a magnified view of the entire mechanical scheme (shown in Fig. 6) focused on the part of the astronomical model.

Fig. 22: The mechanism inside the double disc containing a cog wheel having four pins instead of cogs (Photo by Izawa).

such gears, but the original one has them (Fig. 22). The Seiko engineers decided to replace such gears with ordinary gears with 12 cogs. This is another technological feature of the Myriad Year Clock that differs from contemporary horological techniques.

One notable technical feature of this mechanism is that it is not supported by the central axis, but by the two detached axes. The curved sticks for the red and silver balls rotate around the epicentre and cross the centre when rotated. The space between the two discs therefore should be open for their annual rotations. The two discs are accordingly supported by the two independent axes fixed and respectively attached to the concave hemispheric bowl below and the convex hemispheric Earth above. The double disc mechanism itself is driven by a different gear set at the periphery of the concave bowl, which rotates once per day.

The astronomical model has another design feature, which shows the Japanese calendrical and time-keeping system. We see here the view of the replicated astronomical model without glass dome (Fig. 23). It has the circular plane of the Japanese map in the middle and the annular plane outside, between which the solar and lunar balls sink and rise. We must take note that the levels of the two planes—the map and the ring—are slightly

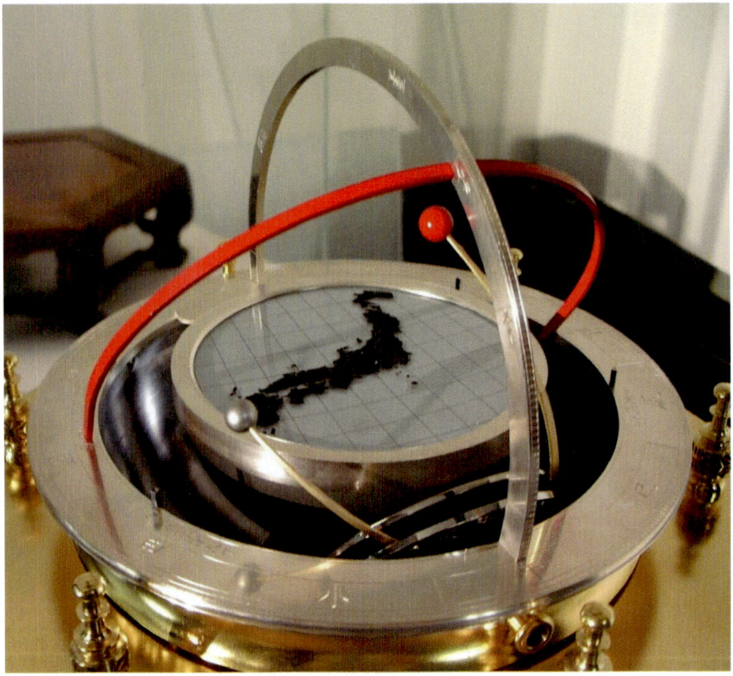

Fig. 23: The astronomical model at the top consisting of two levels. The model displays a Japanese map on the level of the ground and has an outer circular rim. The level of the rim indicates the level below the horizon (Photo by Izawa).

different. The level of the outer ring is set just a little lower than the level of the map. The level of the map represents the horizon, whereas the level of the outer ring represents the position a certain degree below the horizon. The amount of this angle is geometrically determined from the calendrical system at the time. It defines the beginning and end of the day as the time of twilight, or more specifically and scientifically the time when the sun is 7 degrees and 22 minutes below the horizon. This is the time, at the two equinoxes, 36 minutes (or 2.5/100ths of a day) before the sunrise or after the sunset. This definition of the twilight time is important because the time-keeping system in Edo Japan was the seasonal time system, and one Japanese hour, or one *shinkoku*, was determined by the length of the daytime, which was in turn determined by the exact beginning and end of the daytime. This important angle of 7 degrees and 22 minutes is visually indicated by the difference between the level of the map and the outer ring. Tanaka Hisashige's advertisement of the Myriad Year Clock contains a picture showing the construction of the astronomical model (Fig. 24a). It clearly shows the difference between the levels of the map and the outer ring. His explanatory note states that the indicated positions at the outer ring are where the sun sets at the solstices and equinoxes and calls the outer ring 'the boundary of dawn and dusk' (晨昏際) (Fig. 24b).

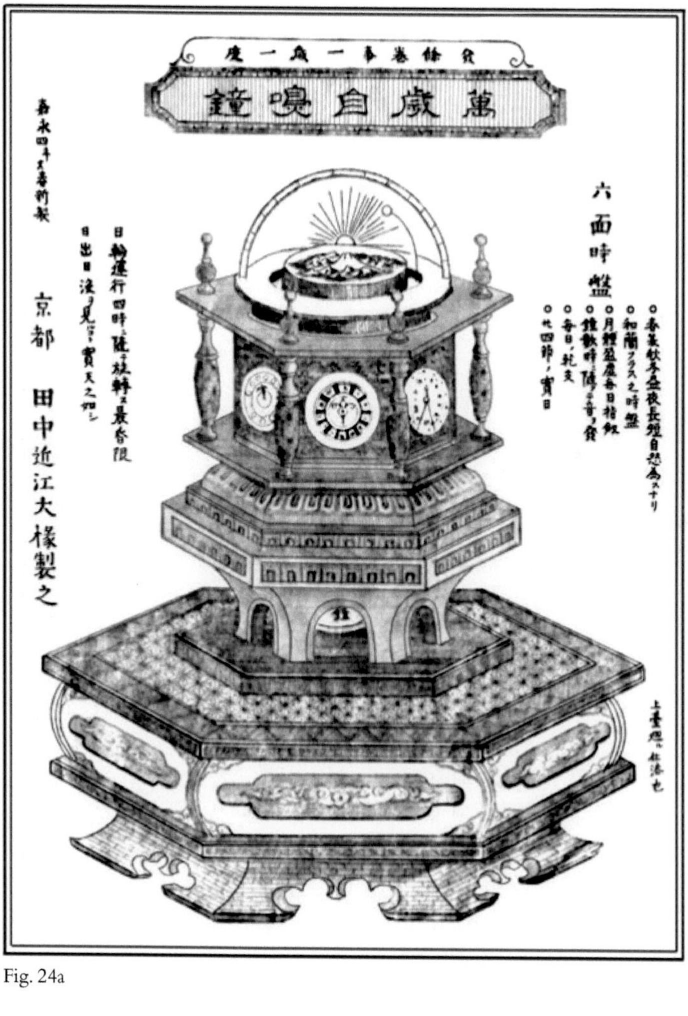

Fig. 24a

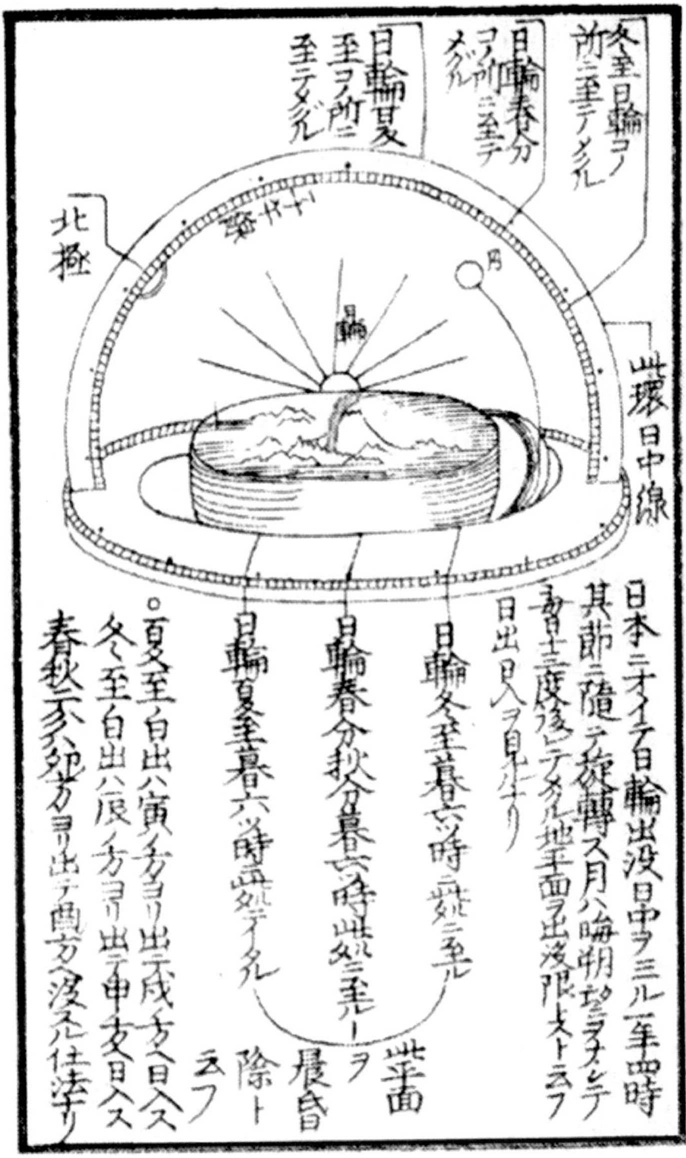

Fig. 24b: The advertisement of the Myriad Year Clock (24a) and the explanation of twilight time in it (24b) (Tanaka Hisashige, Manzai Jimeishō (The Myriad Year Clock), Kyōto, 1851).

8. CONCLUSION

I would like to summarize the technological characteristics of Tanaka Hisashige's Myriad Year Clock. For those of us living in the 21st century, the Myriad Year Clock would not appear to be particularly fantastic on first glance. But once we peek inside, we will be amazed by the complex and intricate mechanisms composed of various parts, from tiny screws to heavy plate coils. The machine is 38 kg and 60 cm in height and consists of more than 1,000 mechanical parts. What you should take away from this is Hisashige's ability as an engineer to plan and design the entire structure as well as the mechanical details of this large mechanical device.

The Myriad Year Clock is not a purely domestic engineering product in Edo Japan, since Tanaka Hisashige used a Western watch as a time-keeper for the entire system, as well as gear wheels of Western clocks. But his capability to design this complex machine system is truly amazing. Especially the two blocks composing the system—the astronomical model and the Japanese *warikoma*-type clock—are intricately designed and constructed. Tanaka Hisashige made a truly original design for the *warikoma*-type clock, encasing all the mechanisms inside, and solved mechanical puzzles to attain complicated motions of the solar and lunar balls. To make them, he studied calendrical and astronomical theories from scholarly experts in these fields. It should be pointed out, however, that the mechanical complexity of these two blocks was a weak point of the Myriad Year Clock because it causes difficulty in keeping the entire system moving by the large torque of the heavy spring coils.

Tanaka Hisashige did not use machine tools, except perhaps a simple instrument to carve screw threads. That would be the most notable technological feature that distinguishes his machine from all other contemporary machines in the West. Owing to its handmade nature, he had to divide angles with his eyes and hands, cut and file the cogs of gear wheels, and adjust the position of the hole for a gear shaft. As microscopic observations show, his cutting technique allowed him to make cogs and wheels to the accuracy of a sub-millimetre.

Lastly, I would like to tell the rest of the story about the Myriad Year Clock and Tanaka Hisashige himself. He intended to sell the clock, even making an advertisement. It is said that the Lord of the Izumo domain (present Shimane prefecture) was intrigued by the clock and considered purchasing it. His bureaucrats, however, advised that it would be too expensive. It was not purchased by the lord or any other customer, but it attracted the attention of those who knew the importance of engineering. Just two years later, Hisashige was invited by the Saga domain to work on the construction of steam-powered machines, small-scale models at first and real-size machines later.[19] It is certain that Tanaka Hisashige's capacity to design large and complex mechanisms as well as make parts and machines with sub-millimetre precision contributed to the success of Saga domain's early attempt at modernization. It is also certain that his experience at Saga as well as his long-time experience as an inventor contributed to subsequent engineering activities after the Meiji Restoration, eventually leading to the establishment of Toshiba Company.

NOTE ON PHOTOGRAPHS

During the project of the investigation and replication of the Myriad Year Clock, numerous photographs were taken by participants, mostly by the two members of the Seiko group, Izawa Minoru and Yamakoshi Toshio, and by the present author. Izawa mainly took photos of components of the original and the replicated Myriad Year Clock as well as the process of disassembling and constructing the components, and all the photos were classified according to the blocks of the Myriad Year Clock. Yamakoshi mainly took photos of the scenes of the mechanical parts and technical work during the project, and the photos were classified according to the dates of the work.

REFERENCES

Asahina Teiichi and Oda Sachiko, ''Myriad Year Clock' Made by G.H. Tanaka 100 Years Ago in Japan', in *Kokuritsu kagaku hakubutsukan kenkyū hōkoku*, no. 35, 1954.

Frumer, Yulia, *Making Time: Astronomical Time Measurement in Tokugawa Japan*, Chicago: University of Chicago Press, 2018.

Hashimoto Manpei, *Nihon no jikoku seido* (Time system of Japan), Tōkyō: Hanawa Shobō, 1994.

Hashimoto Takehiko, 'Kanseireki to wadokei: Yoake no teigi o megutte' (Kansei calendar and Japanese clocks: concerning the definition of twilight), in *Tenmon Geppo*, vol. 98, 2005, pp. 373–379.

---

19 Tanaka Ōmi Ō Kenshōkai (eds), *Tanaka Ōmi Daijō*, Tōkyō, 1931, pp. 76–126; Ōhashi Shūji, *Bakumatsu Meiji seitetsu-shi* (A history of iron making at the end of the Edo and Meiji periods), Tōkyō: Agune-sha, 1975, p. 55.

Hashimoto Takehiko (ed.), 'Tokei no gijutsuteki tokuchō to shakiteki igi ni kansuru rekishiteki kenkyū' (Historical Researches on Technological Characteristics and Social Significance of the Myriad Year Clock), unpublished report of the Grant-in-Aid for Scientific Research of the JSPS (Japan Society for the Promotion of Science), 2006. See the catalogue of the National Diet Library, Tōkyō, https://ndlonline.ndl.go.jp/#!/detail/R300000001-I000008447651-00.

Hashimoto Takehiko, 'Mannen dokei fukugen, fukusei purojekuto to bunkai chōsa' (The project to restore and replicate the Myriad Year Clock and the dissambling investigation), chapter 3 of the JSPS report. 2006, pp. 30–38.

Hashimoto Takehiko, 'Mannen dokei kikōbu no bunkai chōsa hōkoku' (A report of disassembling investigation of the Myriad Year Clock), chapter 5 of the JSPS report. 2006, pp. 51–112.

Hashimoto Takehiko, 'Mechanization of Time and Calendar: Tanaka Hisashige's Myriad Year Clock and Cosmological Model', *UTCP Bulletin*, vol. 6, 2006, pp. 47–55.

Hashimoto Takehiko, 'The Adoption and Adaptation of Mechanical Clocks in Japan', in Feza Günergun and Dhruv Raina (eds), *Science between Europe and Asia: Historical Studies on the Transmission, Adoption, and Adaptation of Knowledge*, Dordrecht: Springer, 2011, pp. 137–149.

Hato Takehiro, Suzuki Kazuyoshi, Tomii Yoichi, Yoshida Mitsunobu, Yokota Yasuhiro, and Kubota Yuji, 'Mechanism of 'Mannen dokei', a Historic Perpetual Chronometer (2nd Report, Power Supply)', *Transactions of the JSME (in Japanese)*, vol. 73, no. 729, 2007, pp. 1537–1544.

Imatsu (Imazu) Kenji, *Karakuri Giemon: Tōshiba sōritsusha Tanaka Hisashige to sono jidai* (The Ingenious Giemon: The Life and Times of Hisashige Tanaka, the founder of Toshiba), Tōkyō: Daiyamondo-sha, 1992.

Kinoshita Yasuhiro, Mannen Dokei no bunkai chōsa kara mita Bakumatsu seiyō gijutsu juyō ni kansuru kenkyū: Mannen Dokei no gijutsuteki genkai to Tanaka Hisashige no gijutsusha e no tenkan (A research on the acceptance of western technologies at the end of the Edo period seen from the disassembling investigation of the Myriad Year Clock: the technological limitation of the Myriad Year Clock and the transformation of Tanaka Hisashige into an engineer), Ph.D. dissertation, Tōkyō Institute of Technology, 2015.

Mody, N.H.N., *Japanese Clocks*, Rutland, VT and Tōkyō: Charles E. Tuttle, 1932.

Oda Sachiko (ed.), *Seiko Tokei Shiryōkan to wadokei roku* (A record of Japanese clocks preserved in Seiko Clock Museum), Tōkyō: Seiko Tokei Shiryōkan, 1986.

Ōhashi Shūji, *Bakumatsu Meiji seitetsu-shi* (A History of Iron Making at the End of the Edo and Meiji Periods), Tōkyō: Agune-sha, 1975.

Tanaka Hisashige, *Manzai Jimeishō* (The Myriad Year Clock), Kyōto, 1851.

Tanaka Ōmi Ō Kenshōkai (eds), *Tanaka Ōmi Daijō* , Tōkyō, 1931.

Tsuchiya Hideo, 'Tanaka Hisashige 'Mannen Dokei' chōsa hōkoku' (A report on the investigation of 'the Myriad Year Clock' of Tanaka Hisashige), *Micromechatronics*, vol. 50, no. 194, 2008, pp. 58–68.

Yamaguchi Ryūji, *Nihon no tokei* (Clocks in Japan), Tōkyō: Nihon hyō ron-sha, 1950.

Yokota Yasuhiro, Suzuki Kazuyoshi, Yoshida Mitsunobu, Hato Takehiro and Kubota Yuji, 'Mechanism of 'Mannen dokei', a Historic Perpetual Chronometer (1st Report, Celestial Globe and Japanese Traditional Clock)', *Transactions of the JSME (in Japanese)*, vol. 73, no. 729 (2007): pp. 1529–1536.

6

A Statistical Analysis of *Tōkyō Meikō Kagami* (with a Focus on Highly Skilled Metalwork Craftsmen)

NISHIYAMA Takahiro

1. INTRODUCTION

TŌKYŌ MEIKŌ KAGAMI is a two-volume catalogue of highly skilled craftsmen (*meikō*) in the Tōkyō area (Fig. 1).[1] It was published in 1879 and contains the names, ages, and residential areas of workers in craft firms, as well as the craft's field of production, its number of workers, and the history of the firm's establishment. The publication was considered to be a highly informative source for exploring craftsmen and craft works, as well as their career formation. The method of surveying was interviews with a precompiled questionnaire. The two volumes provide a summary of these interviews.

The title of the volumes, *Tōkyō meikō kagami*, contains three words: the place name Tōkyō; *meikō*, meaning highly skilled (literally "excellent") craftsmen; and *kagami*, meaning "metal mirror." The metal mirror metaphorically suggests successful model examples. The mirror reflects the actual state of one's form and forces one to recognize the conditions that must be improved and mastered. *Tōkyō meikō kagami* thus represents the models provided by the highly skilled craftsmen resident in Tōkyō – in other words, "model examples of highly skilled craftsmen in Tōkyō".

A total of 650 craftsmen who met the selection criteria in the Tōkyō area were interviewed. Only one was a woman, a jew-

[1] Tōkyō-fu kangyō-ka, *Tōkyō meikō kagami* (2 vols.), Tōkyō: Yūrindō, 1879.

明治十二年十二月印行

東京名工鑑

乾

東京府勧業課

Fig. 1: Title page of the book *Tōkyō meikō kagami* (Model examples of highly skilled craftsmen in Tōkyō). Source: Tōkyō-fu kangyō-ka (ed.), *Tōkyō meikō kagami* (2 vols.), Tōkyō: Yūrindō, 1879.

ellery manufacturer who had completed her apprenticeship in Austria. Of these 650, I focus here on 106 metalworking craftsmen, utilizing the definition by Odaka.[2] My sample included 22 foundry workers (*imono-kō*), 27 blacksmiths (*kaji-kō*), 31 precious-metal ornament workers (*kazari-kō*), two tinsmiths, four electrical equipment manufacturers, 13 medical-instrument makers, two craftsmen who were familiar with the technology of the foundry (*kinzoku chōkō*), and three mechanics who manufactured fine equipment (*zatsu-kōshō*).

A survey carried out by the government in Tōkyō-*fu* in 1877, two years before the publication of *Tōkyō meikō kagami*, indicates that the total number of craftsmen resident in Tōkyō was

[2] Odaka Kōnosuke, *Shokunin no sekai, kōjō no sekai* (The world of craftsmen and the world of factory), Tōkyō: Libroport, 1993.

73,843.[3] With respect to metalwork craftsmen, the total number resident in the Tōkyō area was 3,546.[4] Accordingly, the proportion of highly skilled metalworkers registered in *Tōkyō meikō kagami* was low. For this reason, the results of the following statistical analysis do not present an overall picture. However, they do provide insight into metalworking craftsmen of the transitional phase of 'Japan's industrial apprenticeship.'[5]

The era of Japan's industrial apprenticeship, an era of technological change driven by a wave of technology imported mostly from foreign countries, was accompanied by the substitution of basic materials, particularly in the production of heavy equipment such as ships or cannons, from wood or copper to iron. The Tokugawa shogunate, which aimed for rapid militarization of the country, first tried to combine traditional craft knowledge with western technical knowledge. The government had hoped for an efficient transfer of knowledge from the West.

In this paper, I examine whether there was a sufficient "critical mass" of knowledge carriers, whether traditional craft knowledge was generally applicable for new westernized metalworking, and whether a knowledge transfer from the West was ongoing at the craftsman level.

First, I present the overall picture of the *meikō*-craftsmen in the Tōkyō area. Then I discuss the particularities of the metalworkers, who dealt with iron in the early days of Japan's industrialization. Thereafter, I focus on the careers and vocational training and knowledge transfer from the West to the skilled metalwork craftsmen in Tōkyō. Thereafter, I describe the structure of firms, such as the number of employees and apprentices, as relevant to my research interest. Finally, I summarize the results of the survey.

2. GENERAL INFORMATION ABOUT THE *TŌKYŌ MEIKŌ KAGAMI*

The number of professions among the registered highly skilled craftsmen in *Tōkyō meikō kagami* was 66. Of the ten most popular professions among the *meikō*-craftsmen, five were in the category

3 Tōkyō-fu, *Tōkyō-fu tōkei-hyō, Meiji 10nen* (Statistics of the Tōkyō prefecture, Meiji 10), Tōkyō, 1882.

4 Tōkyō-fu, *Tōkyō-fu tōkei-hyō, Meiji 10nen*, pp. 184–185.

5 This term was coined by Erich Pauer, "Technologietransfer und industrielle Revolution in Japan 1850–1920" (Technology Transfer and Industrial Revolution in Japan 1850–1920), in *Technikgeschichte*, Vol. 51 (1984), No. 1, pp. 34–54.

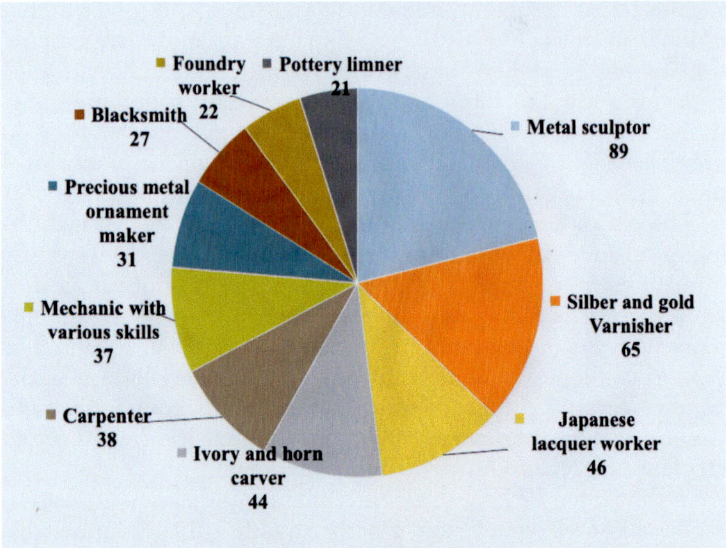

Fig. 2: The ten most frequent professions in *Tōkyō meikō kagami* (Number of persons)

of metalworking: metal sculptor (*kinzoku chōkō*), precious-metal ornaments worker (*kazari-kō*), blacksmith (*kaji-kō*), foundry men (*imono-kō*), and craftsmen with various skills (*zatsu-kōshō*) (Fig. 2). These professions can be roughly divided into two groups. The first was the occupational group that worked on the metal material itself; the second was the professional group that processed the surface of the metal. The first included blacksmiths (*kaji-kō*) and foundry men (*imono-kō*).

The relatively large number of foundry workers in Tōkyō was related to the production of cooking appliances and agricultural equipment in this area until the mid-nineteenth century. The concentration of highly skilled blacksmiths in Tōkyō was, according to Odaka, a consequence of political instability and the increased demand for armaments before and after the Meiji Restoration.[6]

The metal sculptors (*kinzoku chōkō*) and precious metal ornament workers (*kazari-kō*) and a pair of craftsmen with various skills (*zatsu-kōshō*) belonged to the second group. These precious-metal ornament workers made small ornaments for furniture, such as doorknobs and drawer pulls. The craftsmen with various skills

6 Odaka Kōnosuke, *Shokunin no sekai, kōjō no sekai*, passim.

(*zatsu-kōshō*) produced various products. Because they offered a diverse assortment of goods, these workers cannot easily be classified into a specific professional category.

The living areas of all registered *meikō*-craftsmen were mainly Asakusa, Kanda, Nihonbashi, Kyōbashi, Shiba and Honjo (see Figs. 3 and 4). More than half of the selected highly skilled craftsmen in Tōkyō – to be exact, 369 people – lived in these districts. These districts lay along the Sumida River from the northernmost to the southernmost parts of Tōkyō prefecture (Tōkyō-*fu*).

Fig. 3: 15 Districts in Tōkyō in the Meiji era (1886).

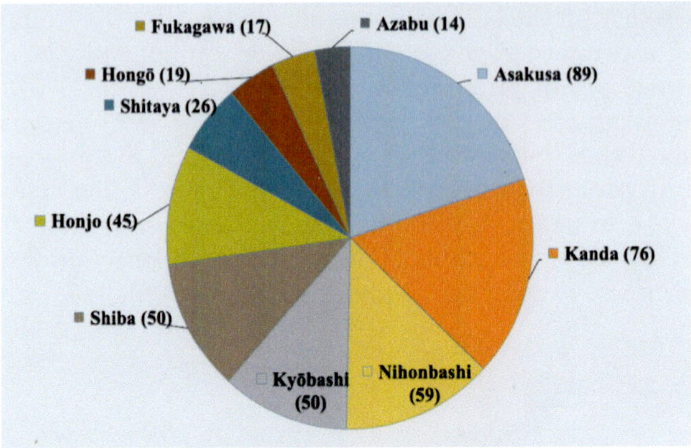

Fig. 4: The ten districts in Tōkyō where most of the highly-skilled craftsmen lived. (Number of persons; calculations by the author).

The concentration of *meikō*-craftsmen living near the Sumida River was related to the fact that this locality allowed for efficient transport over waterways. For this reason, the population in Tōkyō was initially concentrated in these districts.

Most of the highly skilled metalworkers lived and worked in the districts spanning from Kyōbashi to Kanda and in the so-called *shitamachi* (old town) area, Honjō and Asakusa. The highly skilled foundry workers were mostly located in Kanda, Kyōbashi, Honjō, Shiba and Asakusa (Fig. 5). These districts together formed a

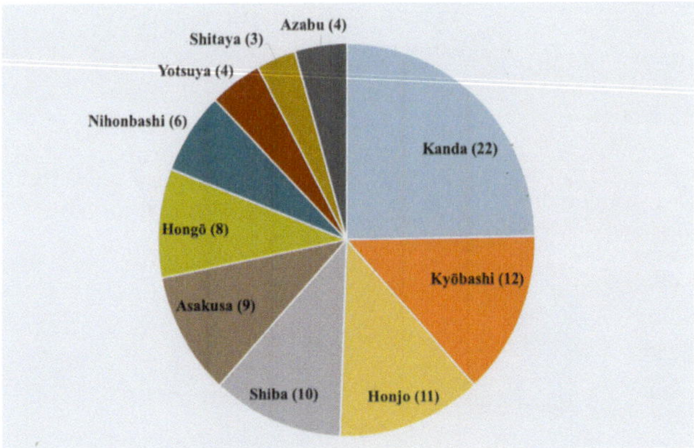

Fig. 5: The ten districts in Tōkyō, where most of the metalworking highly-skilled craftsmen lived (Number of persons: 89 of 106). (Calculations by the author).

triangle in the south-eastern part of Tōkyō. Unlike most *meikō* -craftsmen, who lived along the river Sumida, the foundry workers lived more in the inner-city side of the districts.

Most of the 27 blacksmiths (*kaji-kō*) including two mechanics (*kikai-kō*) lived in Kyōbashi, Nihonbashi and Hongō. The districts of Kyōbashi and Nihonbashi are located directly at the mouth of the Sumida River on Ishikawajima Island (Fig. 6).

In 1853 the Mito domain founded the Ishikawajima shipyard on Ishikawa Island, where the first screw corvette was made entirely by Japanese hands. Ono Tomogorō designed this screw corvette and Hida Hamagorō designed its steam engine. Both of them acquired their technical knowledge at Nagasaki Naval Training Centre (Nagasaki kaigun denshūjo). The Ishikawajima shipyard was privatized in 1876 and in 1888 it began to build iron ships as the first private shipyard in Japan. The presence of this shipyard was one of the reasons for the concentration of smiths in these districts in the late 1870s.

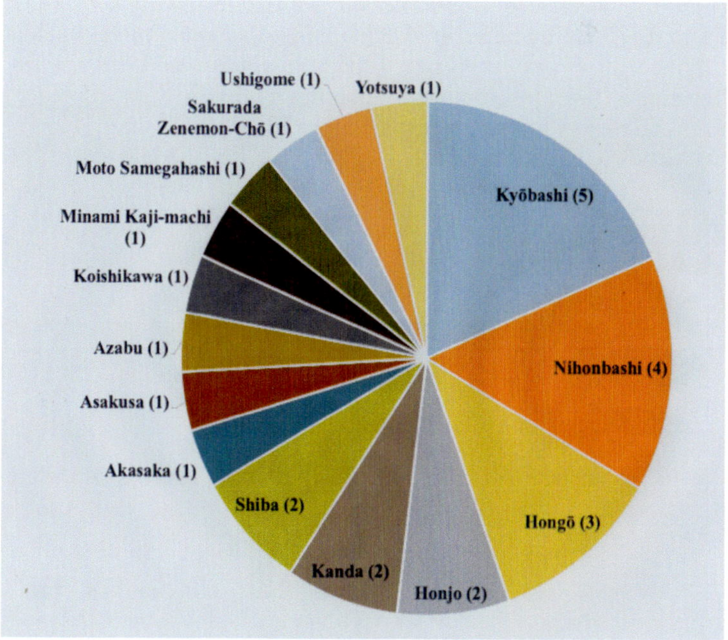

Fig. 6: Districts in Tōkyō frequently inhabited by blacksmiths (total 27). (Calculations by the author).

3. AGE AND APPRENTICESHIP

The average age of the highly skilled craftsmen recorded in *Tōkyō meikō kagami* was 42 years (valid 639, missing 11). Most of these workers were between 26 and 45 years old (Figure 7). The average age of metalworking craftsmen was 43 years (valid 104, missing 2); many of them were between 40 and 44 years old (Fig. 7). However, the younger age group (29 to 39 years) formed an equally large cohort among highly skilled metalworkers. This means that the excellent skills were by no means reserved for older workers.

The medical instrument maker Suzuki Torakichi was the fifth youngest among the *meiko*-craftsmen. He produced portable medical instruments for surgical and ophthalmic devices. The oldest craftsman was an 80-year-old mechanic (*kikai-kō*) named Tanaka Hisashige, who was also known as the *'Karakuri Giemon'* (Giemon, the mechanic). He had invented and built various types of traditional Japanese automata (*karakuri ningyō*) from the 1820s on, and constructed Japan's first planetarium according to a model of the Buddhist universe in 1850 and the Myriad Year Clock in 1851. From 1875 he manufactured telegraph equipment in his Tanaka Engineering Works in Ginza district; his customer here was the Telegraph Bureau of the Meiji government. He had 20 employees in his factory. After the death of Tanaka Hisashige in 1881, his son Daikichi established the Tanaka-seisakujo. After the transfer of the

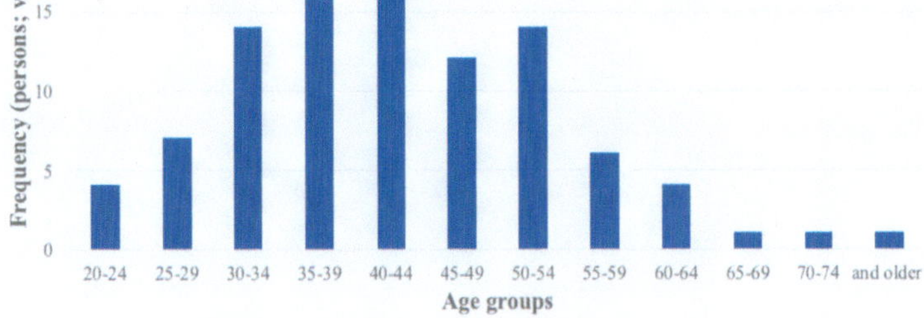

Fig. 7: Age of metalworking craftsmen in *Tōkyō meikō kagami*. (Calculations by the author)

firm to the Mitsui-Zaibatsu in 1893, the name of the company was changed to Shibaura Engineering Works. In 1939 it merged with Tōkyō Denki and became Tōkyō Shibaura Denki, later Tōshiba.

In general the highly skilled craftsmen (*meikō*) were trained in an apprenticeship system. The apprenticeship at a master (*shishō*) took about ten years in all craft industries. The knowledge transfer in the craftsmanship consisted of the acquisition of implicit knowledge, which was incorporated from their master and occurred mainly through the "looking at and learning by doing" principle.[7]

Regarding the distribution of vocational training types, half of the 106 metalworkers known as highly skilled craftsmen had been vocationally trained outside their own family business. Of them, 37% had acquired their exceptionally high skills at home from their parents (Figure 8), and 6 % had acquired their skills autodidactically.

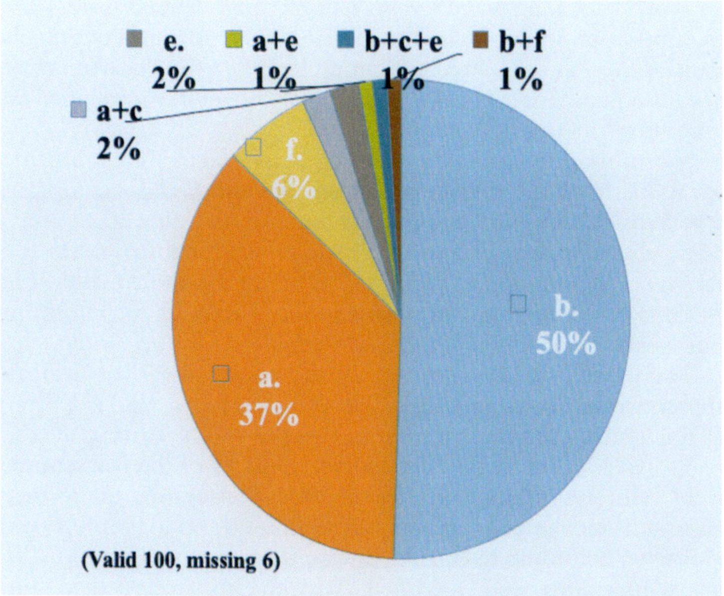

Fig. 8: Distribution of apprenticeship types among metalworkers: a Apprenticeship at family business; b Apprenticeship outside the family business; c Itinerant craftsmen (*watari shokunin*); d In-house training in a company; e Skills taught by European or American; f Autodidactilly. (Calculations by the author).

---

[7] Endō Motoo, *Kinsei shokunin no sekai* (The world of craftsman in the Edo period), vol. III, Tōkyō:Yūsankaku-Shuppan, 1985.

The highly skilled metalworkers often started their business after being apprenticed outside of their home. In three cases, the apprentice had also further trained as itinerant craftsmen (*watari shokunin*) under a guild craftsman. These were the blacksmith Tsunanabe Seikichi, the tinsmith Yoshimoto Ryogen, and the precision mechanic Fujishima Tsuneyoshi. A brief overview of the careers of these three highly skilled metalworkers is provided below.

Tsunanabe Seikichi manufactured small metal devices. He had been trained as a blacksmith at home under his father's guidance and then worked as an itinerant craftsman for five years. When he was 25 years old, he established his own craft business. He received several orders from the Telegraph Bureau, the Army and the universities. His products were lathes, small metal tools for Western furniture, small metal devices for telegraph appliances and rice cookie moulds.

Yoshimoto Ryōgen completed his apprenticeship with a pewter worker until he was 19 years old. He then worked from 1837 to 1850 as an itinerant craftsman. After this itinerant craftsman's training, he was appointed as a group leader in the department of pewter equipment at the Tokugawa government. He developed an innovative tin-processing method.

Fujishima Tsuneyoshi first trained as a sculptor. After his first vocational training, he worked as an itinerant craftsman for half a year in Nagasaki and Saga under a gunsmith. After his training, he taught himself how to manufacture measuring instruments. He also received training in this field from an Austrian master. He was one of the few metal craftsmen who learned their skills from Europeans.

Twenty-two of the 106 highly skilled metalworkers had an apprenticeship as a gunsmith. One of the electric workers, three of the medical instrument makers, three of the measuring instrument makers, four of foundry workers, and 11 of the blacksmiths were trained as gunsmiths. After the Meiji Restoration the former gunsmiths recognized that their profession was no longer relevant. However, according to Suzuki, due to their iron-processing skills, many gunsmiths were recruited as mechanics around that time and were employed in the rising state-owned and state-operated heavy industries.[8]

[8] Suzuki Jun, *Meiji no kikai kōgyō, sono seisei to tenkai* (The machine industry in the Meiji period – its origin and development), Tōkyō: Minerva shobō, 1996, p. 34

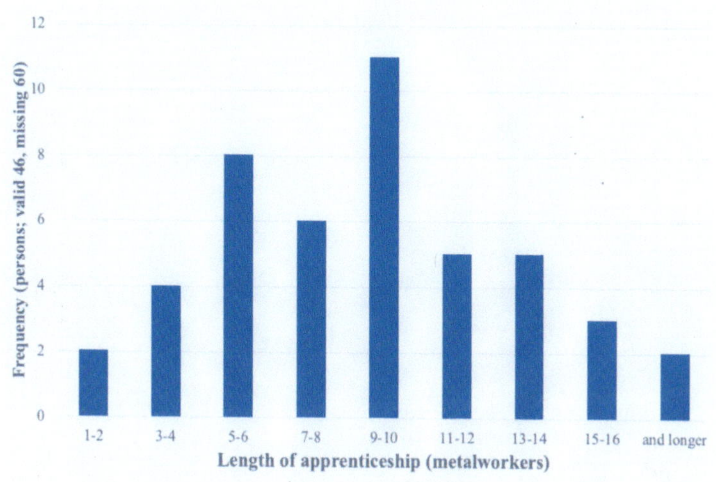

Fig. 9: Length of apprenticeship for highly skilled metalworkers. (Calculations by the author).

The average length of apprenticeship for highly skilled metalworkers was 9.4 years (valid 46, missing 60). This was similar to the average length of apprenticeship across all occupational categories in *Tōkyō meikō kagami*. Some had a shorter apprenticeship of five or six years; these were often metalworkers without an artistic orientation. Three medical instrument makers, two blacksmiths and one foundry worker belonged to this group. The precious-metal ornament workers (*kazari-kō*), who often produced vases with sculptures, swordsticks, glasses or wristwatches and so on, had a much longer training period than blacksmiths and foundry workers who worked with raw or cast iron (Fig. 9)

The average age at which metalwork craftsmen started their vocational training was 15.6 years (valid 46, missing 60). The youngest was a foundry worker aged nine, when he began his training. At the age of 22, he started his own business. This indicates that taking on an apprentice by a master (*shishō*) was similar to adopting a child. A good relationship between the parents and the master was probably a prerequisite for admission to the apprenticeship. Regarding the starting age for an apprenticeship across the occupational categories, the medical instrument makers began their training relatively late (Fig. 10).

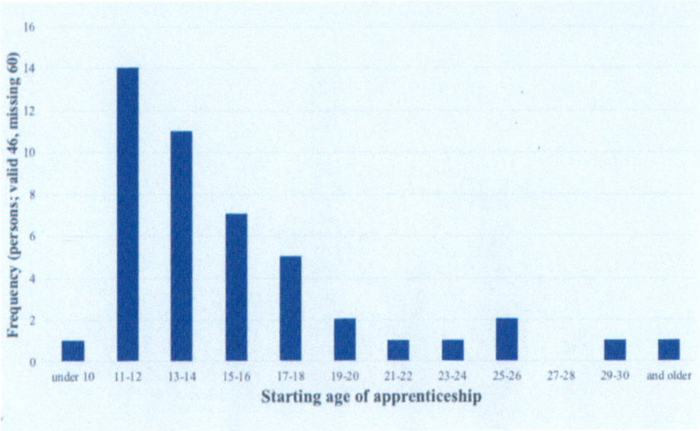

Fig. 10: Age of starting apprenticeship of highly skilled metalworkers. (Calculations by the author).

4. APPRENTICESHIP AND WESTERN TECHNOLOGIES

No significant correlation was found between the entry age and the duration of training. However, the training period of the *meikō* -craftsmen who received their apprenticeship from Europeans or Americans was markedly shorter. Some completed their apprenticeship abroad or under an American or European person residing

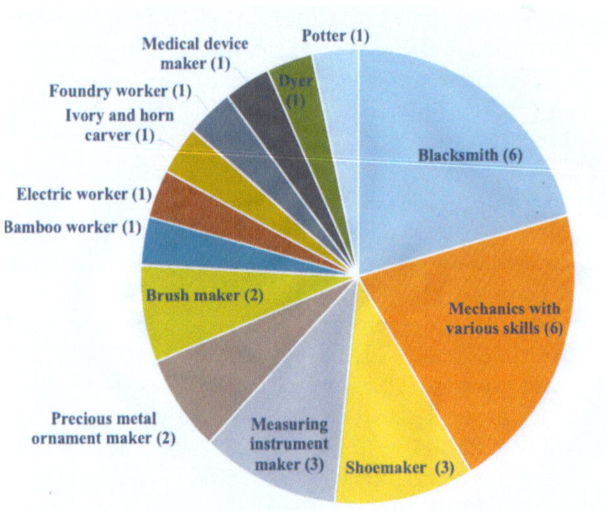

Fig. 11: Number of highly skilled craftsmen influenced by Western technologies (Number of persons; total 29). (Calculations by the author)

in Japan. However, of the 650 *meikō*-craftsmen, only a small number – 29 craftsmen (4.2 %) – had acquired their technical knowledge either from European or American professionals in Japan or through apprenticeship in a Western country (Table 1). They belonged to an elite group selected by the government, and had received scholarships from the state for a study trip. In addition, the existence of craftsmen who had trained autodidactically, yet reached the level of highly skilled *meikō*-craftsmen must be noted (Fig. 11).

Among these 29 *meikō*-craftsmen, the occupational categories of blacksmith (*kaji-kō*), craftsman with various skills (*zatsu-kōshō*), bootmaker (*kutsu-kō*) and manufacturer of surveying equipment (*sokki-kō*) were most strongly influenced by western sciences and technical professionals. Thirteen of the 29 craftsmen had received training directly from western engineers or specialists in Europe or the United States (Table 1). Of these, four had learnt how to manufacture mechanical equipment. Three other metalworkers had completed their apprenticeships under Europeans or Americans. The rest worked as craftsmen with various skills (*zatsu-kōshō*) and provided various goods, such as finger rings, eyeglasses or stuffed animals; to produce these, European knowledge and technology were indispensable.

Name	**Type**	**Profession**	**Length (years)**	**Acquired knowledge**	**Origin of the knowledge**
Kashima Kinsuke	e	blacksmith	4	medicine	The Netherlands/ Germany
Okada Koheiji	a+e	blacksmith	1	blacksmith	Great Britain
Wada Sadaichirō	e	medical device maker	3	blacksmith	Great Britain
Fujishima Tsuneyoshi	b+e	measuring instrument maker		production of measuring instruments	Austria
Ida Rokuzō	e	shoemaker	2	shoemaking	China/The Netherlands
Itō Hiroshi	e	shoemaker	1	shoemaking	The Netherlands
Okamoto Tōzō	e	shoemaker	4	shoemaking	The Netherlands
Sonobe Hideyoshi	e	mechanic	7	chemistry	Great Britain
Asakura Sayo	a+e	mechanic	1	jeweller	Austria
Oda Nobutoku	a+e	mechanic		zoology & botany	Great Britain
Yamaoka Jirō	b+e	dyer	4	chemistry	USA
Yaguchi Tanzan	e+f	mechanic	0.5	chemistry	The Netherlands
Nōtomi Kaijirō	a+b+e	potter	2	pottery	China/USA/Austria (Elbogen in Bohemia)

Table 1: Highly skilled craftsmen, apprenticed under European or American engineers and professionals (type of apprenticeship: e = exclusively under European professionals; a + e = first in family business, then under European professionals; b + e = first outside the family business, then under European professionals) names that (calculations by the author).

For example:

Kashima Kinsuke completed an apprenticeship with a blacksmith in his early professional life. When he was 38 years old, he studied medicine for two years under Koenraad Wolter Gratama, a Dutch chemist, later considered to be the father of Japanese chemistry and, from the age of 40, another two years under a German named Hermann Ritter, also a chemist. After this vocational training under Europeans, he produced medicines and later medical instruments. He had five apprentices working in his factory.

Wada Sadaichirō had his first apprenticeship with a gunsmith. He received three years of training from an English mechanic who lived at Yokohama House No. 69. Wada produced medical instruments after his gunsmith training.

Okada Koheiji had his first apprenticeship as a gunsmith in the family business for one year; he then received further vocational training as a machine maker under an Englishman. After this vocational training, he produced various devices, such as pumps, sewing machines and guns. One employee and three apprentices worked in his factory.

Englishmen (four persons) and Dutchmen (three persons) were the most frequent teachers of western technologies. In developing the skills of *meikō*-craftsmen, knowledge transfer from Europe played a major role. Among the four Englishmen, two were device smiths (*kikai-kaji*).

Noteworthy are three *meikō*-craftsmen who visited Western countries and acquired their skills there. Among this group was a jeweller, Asakura Sayo, who had her apprenticeship in Austria. She was the only woman selected as a highly skilled crafts(wo)man. In addition to (finger) rings, she made Western eyeglasses. This contributed to the industrialization of Japan because of their function of preserving and enhancing the capability of key figures. She had six apprentices in her company.

Fujishima Tsuneyoshi was a mechanic of surveying instruments. He was an apprentice of a master named *Karafuto* (Kraft?) in Austria. Before studying there, Fujishima had already trained as a gunsmith. In addition to surveying equipment, he produced medical equipment; he had 15 employees in his factory.

These highly skilled craftsmen who had acquired western knowledge hired a number of employees in their factories. This indicates that their products were in relatively high demand at that time. Moreover, the growing number of people who had Western knowledge and trained many apprentices reflects a cycle of steadily advancing dissemination of Western knowledge in Japan.

5. THE OPERATING STRUCTURE OF CRAFT FIRMS UNDER THE HIGHLY SKILLED CRAFTSMEN

In 1870, the Ministry of Public Works (*Kōbushō*) was established by the Meiji government. Its industrial policy was promoted with the slogans *shokusan kōgyō* ("promotion of industry") and *fukoku kyōhei* ("rich nation, strong army"). The industrialization and militarization of Japan was also promoted by this ministry. Many state-owned pilot factories were subsequently established in the fields of railway engineering, mining, silk spinning, shipbuilding and mechanical engineering.[9] Although the modernization of the economy driven by the Japanese government did not directly promote craft firms, it also had a significant impact on small private business owners.

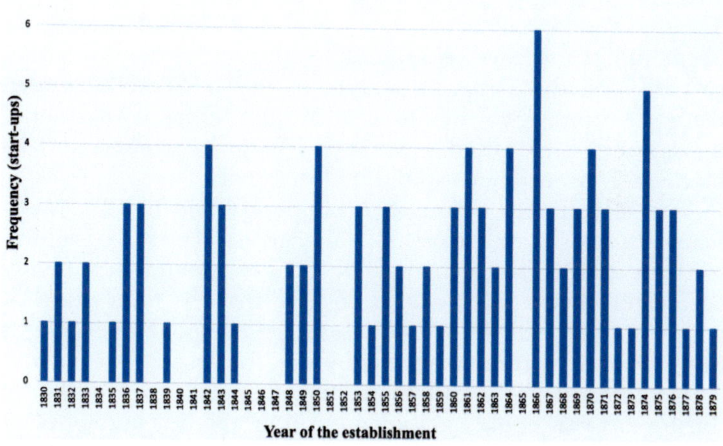

Fig. 12: Number of enterprises and year of establishment by highly skilled metalworking craftsmen between 1830 and 1880 (calculations by the author).

9 Pauer, Erich, 'Japans Aufstieg zur Weltwirtschaftsmacht, Wirtschaftsentwicklung, Wirtschaftsstruktur und Arbeitsbeziehungen' (Japan's rise to world economic power, economic development, economic structure and industrial relations), in *Japan*, Stuttgart: Kohlhammer, pp. 102–131, 1985, esp. p. 104.

Generally, the companies of highly skilled craftsmen were established in the decade between 1866 and 1877. In particular, in 1872, 30 craftsmen founded their businesses (Figure 12).

As mentioned, gunsmiths played a key role as a mediator for the acceptance and takeover of Western technology.[10] For the development of heavy industry through knowledge and technology transfer from the West, craftsmanship formed the basis for this reception, although this alone was insufficient for the new technologies. The metalworkers in the Tōkyō area were considered better than those in Nagasaki, both in their quality of work and in their working speed.[11] For this reason, a number of metalworking craftsmen were recruited into Yokohama Ironworks and Yokosuka shipyard already in 1865 from the Tōkyō area.

Regarding the trends of business start-ups, it is astonishing that in 1865 no metalworking craft business was founded. However, the construction of workshop houses for the Yokohama Ironworks started in March 1865 and ended in October 1865. Hence, 1865 was the year in which the largest ironworks in Japan – with a foundry, blacksmith works, boiler making, woodworking, rigging workshops and so on – were founded in the Tōkyō area and skilled craftsmen were need. The Tokugawa government had provided, as a founding concept for the ironworks, that 100 skilled craftsmen should be recruited for the integrated factory. According to the Director of the Yokohama Ironworks, de Rotour, seven foundry workers, 10 copper workers and 10 blacksmiths from Tōkyō were hired in 1865.[12]

The recruitment of metalworking craftsmen for the Yokohama Ironworks and Yokosuka Shipyard was certainly the reason why in 1865 not a single crafts firm was founded by highly skilled metalworkers. It is also possible that the foundation of craft firms planned for in 1865 had been postponed because of the recruitment announcement by the Tokugawa government. The foundation of a craft firm, for example as a blacksmith, during this year could have been misinterpreted as ignorance of the shogunate's policies. The abrupt increase of business start-ups in 1866 can be explained by the fact that metalworkers

10 Suzuki, *Meiji no kikai kōgyō*, p. 34.

11 Suzuki *Meiji no kikai kōgyō*, p. 12.

12 Etō Atsushi (ed.), *Kaigun Rekishi I* (The History of the Naval Force I), (Katsu Kaishū zenshū, vol. 18), Tōkyō: Kōdansha, 1973, pp. 168–169.

had withheld their start-ups until government recruiting was completed.

Among all 650 highly skilled craftsmen were 208 registered as employer of craftsmen. The average number of employees among them was 5.3 persons (standard deviation: 7.5). (Tab. 2)

With regard to occupational categories, foundries had the most employees as well as the largest number of retained apprentices (Tab. 2). However, the high standard deviation should be noted. Among seven foundries, two firms had many more employees than the others. These were Suzuki Chōkichi in Kyōbashi district, with 25 employees, and Obata Chōgorō in Honjo district, with 16 employees. Suzuki had eight apprentices and altogether 33 employees in his forge. His business produced vases, incense burners and trinkets. The main customer of his products was an export company, Kiritsu kōshō kaisha (Kiritsu Industrial and Commercial Company), which was established by the Meiji government after the Vienna International Exposition in 1874.[13]

Table 2: Average number of employees by occupations of metalworkers

Occupation	Average number of employees	Maximum	Minimum	Number of businesses
Foundry worker (*imono-kō*)	8.1	25	2	7
Measuring instrument maker (*sokki kō*)	8.0	15	1	2
Blacksmith (*kaji-kō*)	4.5	17	1	13
Mechanic (*kikai-kō*)	4.0	5	3	2
Precious metal ornament maker (*kazari-kō*)	3.7	11	1	13
Medical device maker (*iryōki-kō*)	2.8	5	2	5
Mechanics with various skills (*zatsu-kō*)	1.0	1	1	1
Average/Total	4.7	--	--	43

(Total 106, valid 43, missing 63)

13 *Tōkyō-fu kangyō-ka* 1879, vol. 2, p. 64.

Obata, by contrast, produced steam engines and other types of machines. He was initially trained as a gunsmith. From 1873 onwards, he produced mechanical devices. His main customers were the Paper Money Bureau (Shihei-ryō), the Manufacturing Bureau (Seisaku-ryō) and the Ōji Seishi Kaisha (Ōji Paper Co. Ltd.).[14]

6. CONCLUSION

The aim of this article was to highlight some characteristics of the highly skilled metalworking craftsmen in the Tōkyō area during the late Edo- and early Meiji-period. Analysis of the data in *Tōkyō meikō kagami* provided insights into connections between individual craftsmen and the historical events around that time. Particularly noteworthy are the following:

1. Significant correlations exist between mechanical equipment manufacturers and vocational training as a gunsmith.

   Suzuki's hypothesis that many mechanics in factories in the Tōkyō area in the early Meiji period came from the profession of gunsmith applies to my sample, too.[15]

   Among the 49 equipment makers (= mechanics) classified as *meikō*-craftsmen, 22 (about 45%) were former gunsmiths or apprenticed to a gunsmith. Even some joiners had previously completed training as a gunsmith.[16] Well-known and highly skilled mechanics in the *Tōkyō meikō kagami*, such as the inventor Tanaka Hisashige, the small-appliance maker Okada Koheiji, and the medical-instrument maker Wada Sadaichirō, had also completed apprenticeships as gunsmiths.
2. Some knowledge transfer from the West to Japan occurred at the level of craftsmen.

   Among 650 skilled craftsmen, 28 had acquired technical knowledge from the West. Of these, 12 had completed their training directly under foreign technicians or engineers living in Japan or abroad. Many of these highly skilled craftsmen trained more apprentices than the average number and had many employees. This implies the rapid spread of Western knowledge among craftsmen. The knowledge transfer from the West to Japan occurred gradually in the era of Japan's

14 Tōkyō-fu kangyō-ka 1879, vol. 2, pp. 73–74
15 Suzuki, *Meiji no kikai kōgyō*, pp. 34–35
16 Tōkyō-fu kangyō-ka, vol. 2, p. 348

industrial apprenticeship during 1870s, also at the level of craftsmen.

3. The establishment of Yokohama Ironworks and Yokosuka Shipyard resulted in fewer foundations started by blacksmiths and foundry workers and possibly resulted in knowledge transfer from craftsmen to technicians.

   With regard to highly skilled metalworkers, the start-up boom collapsed in 1875 due to the industrialization programme initiated by the Ministry of Public Works. Many highly skilled metalworkers such as foundry men and blacksmiths were recruited into the public corporations of Yokohama Ironworks and Yokosuka Shipyard. The number of *shokkō* (factory craftsmen) employed in the Yokosuka Shipyard in 1876 was over 1,500. The establishment of the Yokohama Ironworks and the Yokosuka Shipyard had a great impact on the reduction of foundations of blacksmiths and foundries in Tōkyō. However, it initiated knowledge transfer from craftsmen to technicians.

From this outline of highly skilled craftsmen in the Tōkyō area, I argue that some played a key role in this era of 'Japan's industrial apprenticeship'. The establishment of the Ministry of Public Works in 1870 and the resulting accelerated industrialization led to the erosion of traditional crafts but also to the emergence of new professions, as well as the need for transformation of certain professions – such as from gunsmith to mechanic. The acquisition of Western knowledge and thus the knowledge transfer from the West to Japan was the key concept behind such career transitions. Furthermore, the absorption of metalworking craftsmen into large enterprises initiated by the new Meiji government shaped craftsmen's integration into the industry. In this regard, *Tōkyō meikō kagami* reflects not only the exemplary nature of the highly skilled craftsmen, but also the epoch of transition from a handicraft to an industrial society in Japan.

REFERENCES

Endō Motoo, *Kinsei shokunin no sekai* (The world of craftsman in the Edo period), vol. III, Tōkyō:Yuzankaku, 1985.

EtōAtsushi (ed.), *Kaigun rekishi I* (The history of the naval force I), (Katsu Kaishū zenshū, vol. 18), Tōkyō: Kōdansha, 1973.

Odaka Kōnosuke, *Shokunin no sekai, kōjō no sekai* (The world of craftsmen and the world of factory), Tōkyō: Libroport, 1993.

Pauer, Erich, 'Technologietransfer und industrielle Revolution in Japan 1850–1920' (Technology transfer and industrial revolution in Japan 1850–1920), in *Technikgeschichte*, vol. 51 (1984), no. 1, pp. 34–54.

Pauer, Erich, 'Japans Aufstieg zur Weltwirtschaftsmacht, Wirtschaftsentwicklung, Wirtschaftsstruktur und Arbeitsbeziehungen' (Japan's rise to world economic power, economic development, economic structure and industrial relations), in *Japan*, Stuttgart: Kohlhammer, 1985, pp. 102–131.

Suzuki Jun, *Meiji no kikai kōgyō, sono seisei to tenkai* (The machine industry in the Meiji period – its origin and development), Tōkyō: Minerva Shobō, 1996.

Tōkyō-fu (eds), *Tōkyō-fu tōkei-hyō, Meiji 10nen* (Statistics of the Tōkyō prefecture, Meiji 10), Tōkyō, 1882.

Tōkyō-fu kangyō-ka (ed.), *Tōkyō meikō kagami* (2 vols.), Tōkyō:Yūrindō, 1879.

7

Boiler Manufacture in Late-nineteenth Century Japan: From First Beginnings to Nationwide Expansion

SUZUKI Jun (Translated by Nicholas Pertwee)

1. INTRODUCTION

TAKING BOILER MANUFACTURING techniques in nineteenth century Japan as its focus, this paper shows what features were considered most beneficial for the country as Western technology was imported, spread and took root, and what part engineers and craftsmen played in this sequence of events.

A boiler is essentially an apparatus formed of a furnace for generating pressurised steam with an accompanying pressure vessel, developed by the Englishman Thomas Savery (1650–1715) shortly before the turn of the eighteenth century. As can be seen from illustrations of boilers of the time, there is a main unit made of copper with a brick cladding, to the bottom part of which heat is applied directly from the fire raised on a grate in which fuel is consumed, with ducting that allows the heated gases thus generated to make one pass round the main body and reheat it *en route* to a chimney. This method of construction assumes knowledge of the European practice of building an oven and a fireplace with bricks.

Steam was known in the Edo period but was only applied to distillation, which meant that boilers were not used in Japan until the middle of the nineteenth century. The art of brick making and bricklaying did not exist until then either, and there was no use made of chimneys. Contemporary ovens made of stones and clay had a common fire hole and smoke vent and the iron cauldron

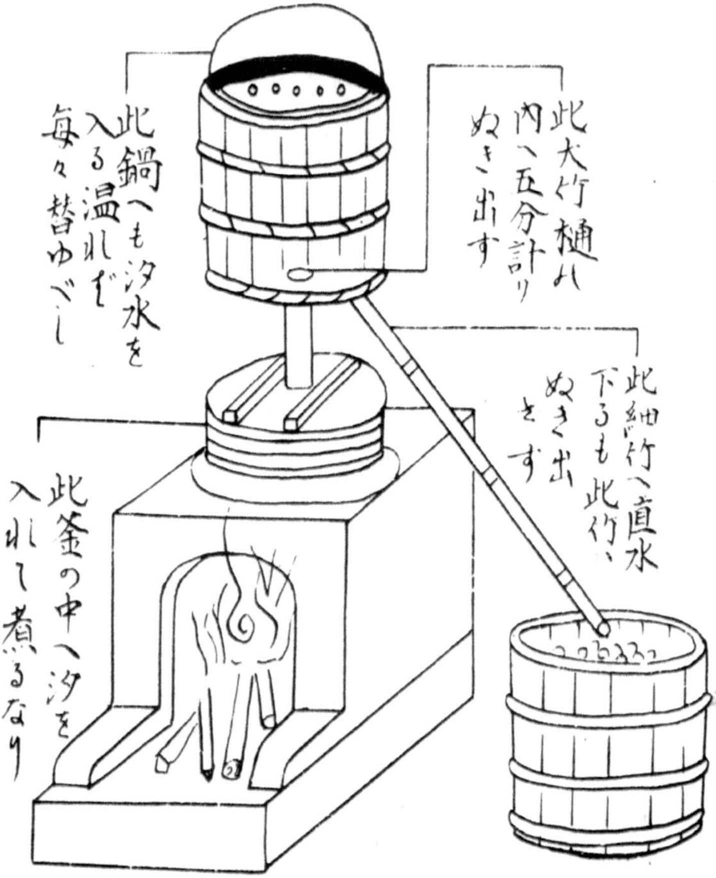

Fig. 1: Distillation apparatus used on-board ship in Japan at the beginning of the 19th century. The clay oven and the iron cauldron inside are ordinary cooking utensils. The distillation tub is placed on top of the oven, and on top of which is a dish filled with sea-water for cooling.

Explanations alongside the drawing.

Top-right: Gases are drawn up into this large bamboo tub for about five minutes

Lower-right: Pure water is drawn off and flows down this narrow bamboo pipe

Top-left: The sea water in this pan has to be changed frequently to regulate the temperature

Lower-left: Sea water is also fed into the oven

Source: Ishiguro Nobuyoshi, *Sanpō tokai hyōteki* (Reference points for sea passages on arithmetical principles), ed. by Suharaya Mohei et al., 1836.

with which they were fitted could only be heated from below (Fig. 1).[1]

By the time the skills needed to make and use boilers were introduced to Japan in the latter part of the nineteenth century boilers had already undergone 150 years of development and the task of designing them based on calculating their strength and efficiency fell to engineers familiar with the theory behind their construction. There were three technical requirements for boilers to be made in Japan: (1) to design them based on theoretical principles, (2) to make their main body out of metal and (3) to make and lay bricks well enough to build enclosing structures. Considerable damage was caused when boilers fractured and a system of regular inspections also had to be in place to prevent such accidents before they could be properly accepted.

2. THE INTRODUCTION OF BOILER TECHNOLOGY TO JAPAN

The first time that most Japanese knew that there was such a thing as a steam engine with its own boiler was when they saw Commodore Matthew Galbraith Perry's fleet of steam-powered warships sail into Edo Bay in 1853, an event that caused them no little consternation.

Despite this there were already some among the *intelligentsia* who knew about steam-driven warships and their potential from reports brought by the Dutch. The advance of the Western powers into Asia, mainly as a result of the Opium Wars of 1840–1842, had attracted the attention of the ruling class, whose mandate for domestic control depended on military power. Saitō Kaoru's (1815–1852) work *Ahen shimatsu* (Opium and its ramifications), which was put into circulation in 1843, spoke of the importance of the English 'steamers' as a fighting force.[2] The Satsuma fiefdom was paying ever closer attention to seaborne military power as it extended its influence to the Ryūkyū islands and in 1848 its chieftain Shimazu Nariakira commissioned *Rangakusha*, Japanese scholars of Dutch matters, to translate Gideon Jan Verdam's

---

1 Ishiguro Nobuyoshi, *Sanpō tokai hyōteki* (Reference points for sea passages on arithmetical principles), ed. by Suharaya Mohei, et al., 1836.

2 Fifth picture in Waseda University Library's Open Pictures http://archive.wul.waseda.ac.jp/kosho/bunko08/bunko08_c0202/bunko08_c0202_p0005.jpg

(1802–1866) *Volledige verhandeling over de stoomwerktuigen* (publ. 1832) and this appeared in the following year as *Mizumushi-sen setsuraku* (A study of steamships). Verdam had studied at Leiden University and later became a professor there. In his book he explained the theory behind the steam engine and gave examples of its use on land, as well as at sea. The Japanese version translated the parts of his book on basic theory and on engines for marine use and, following his assumption of the fief chieftainship in 1851, Shimazu Nariakira ordered the construction of a small marine engine based on the information it contained. The boiler of this marine engine dispensed with bricks and consisted of a main unit made of metal and a chimney, which showed that practitioners of existing metal working skills were involved. Then, in 1855, having seen actual examples of the engines in Dutch steam-driven warships and other, smaller, steamers Shimazu ordered that one be built. The boiler used in it was made of copper plates and was the first working example of a boiler made in Japan.[3] It is a feature of Japan's introduction to boiler technology that its military significance and the theory behind it were discussed before an actual example appeared, which led to the quick response from the Bakufu and some of the domains following Perry's arrival.

In 1855 the Bakufu learnt about naval technology from a Dutch naval squadron at Nagasaki and how to use boilers to operate steam-driven warships. The first fixed land engine in Japan was installed at the engineering works set up at Nagasaki in 1858 as part of this process, and guidance was given in the manufacture and repair of engines incorporating boilers.[4] At the beginning of the 1860s a boiler to power a warship was made at Mietsu in Saga based on the knowledge thus gained, which completed the technology transfer at least as far as the manufacture of marine boilers was concerned.[5]

Hendrik Hardes (1815–1871), the Dutch naval engineer in charge of training at Nagasaki, had Japanese artisans turn their hand to making bricks and showed them how to lay them for the cladding for boilers and for chimneys, this with an eye to fac-

---

3 Kōshaku Shimazu-ke hensansho (ed.), *Sappan kaigunshi* (A history of the Satsuma navy) vol. 1, 1928, pp. 597–609.

4 Sugiyama Kenjirō, 'Nairin shiki jōkisen Sentō-maru ni tsuite' (The Sentō-maru screw steamer), in *Chiba shōdai ronsō*, vol. 40, no. 3, 2002, pp. 41–77, esp. pp. 50, 62–63.

5 Nakanishi Hiroshi, *Nihon kindaika no kiso katei* (The fundamentals of Japanese modernization), vol. 1, Tōkyō: Tōkyō University Publishing Association, 1982, pp. 94–98.

tory-building.[6] This was no doubt possible because Hardes was not a university graduate, but a technician who had a wealth of hands-on factory experience.[7] In 1866 the Bakufu then invited forty-three technical experts from France to the naval dockyard at Yokosuka, in the specializations of shipbuilding, machinery, architecture and civil engineering and began the transfer of skills needed for the construction of the dockyard – including brick-firing – and the manufacture of machinery.[8] The responsibilities among the Frenchmen who were training the Japanese were

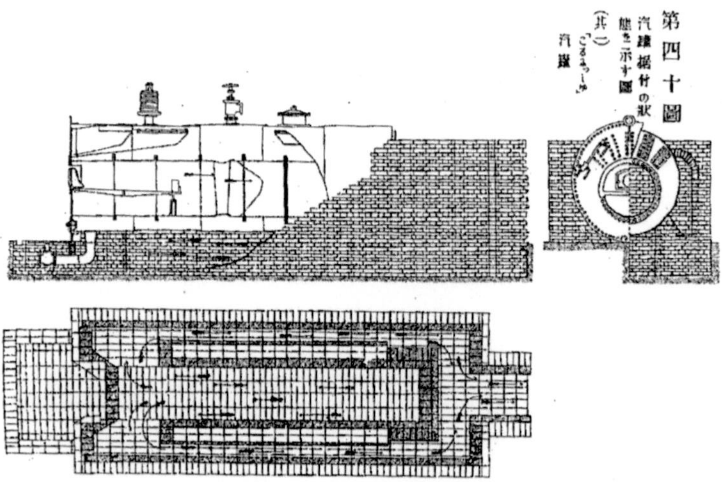

Fig. 2: Cornish boiler with brick cladding: The hot gases produced in the furnace near the flues pass through them to exit on the right-hand side of the boiler. The gases are then redirected under the boiler to the left-hand side via the brick-built smoke duct and pass through the flues on each side of the boiler to emerge once more on the right. From there they are channelled into the chimney at the right which is not shown in this drawing.
Source: Yamamoto Takezō, *Nihon seishi hō* (Silk reeling in Japan), Bunmei-dō, Tō kyō 1909.

6 Kattendyke [Willem Huyssen van Kattendijke], *Nagasaki kaigun denshūsho no hibi* (The story of the Nagasaki Naval Training Centre), translated by Mizuta Nobuyoshi, Tōkyō: Heibonsha (Tōyō Library), 1964, pp. 56,71.

7 Kusumoto Juichi, *Nagasaki seitetsusho* (The Nagasaki Ironworks), Tōkyō: Chūō kōron-sha, (Chūkō shinsho), 1992, pp. 64–65.

8 Yokosuka kaigun kōshō (ed.), *Yokosuka kaigun senshō shi* (The history of the Yokosuka Naval Yard), vol. 1, 1915, pp. 64–77.

probably divided such that university graduates were in charge of designing boilers, while the actual boiler making, selection of the raw materials for bricks, manufacture of bricks and the technique of brick-laying were carried out by specialist craftsmen with those skills. There were chimneys built of bricks at the factories at Nagasaki and Yokosuka as well as boilers with brick cladding (Fig. 2), an indication that the craftsmen who had learnt boiler-making there understood that a brick structure was a requirement for a land-based boiler and who had been given the opportunity to observe the structure of brick-built flues and chimneys. It is also likely that potential buyers who visited these factories to order boilers were able to see how boilers were set up and were able to recruit the bricklayers there for their own factories. When the new government was formed in 1867 foreigners supervised the installation and manufacture of boilers at the factories at Nagasaki and Yokohama run by their compatriots, but also at the numerous government-run factories in Ōsaka and Tōkyō, thus broadening the scope for future dispersal of these activities.

In the 1860s, in an atmosphere of military tension, large numbers of steamships were imported and their operation and repair taken on by Japanese who had been taught by foreigners at Nagasaki and Yokohama, and their compatriots whom they had instructed in turn. This meant that marine boiler technology developed faster than that applicable to land-based steam engines.

3. BOILERS IN THE 1880S

Imaizumi Asuka recently published the results of her research based on the 1889 survey of Tōkyō as the city in which boilers developed most rapidly.[9] Her research showed that forty-one of the seventy-one boilers for which information was obtained were made abroad while among the thirty domestically made ones, at least twelve were second-hand. Then, of the eighteen 'home-made' ones whose manufacturers have been identified, eight came from government factories, two from commercially-run factories that started up after being disposed of by the Government and eight

9 Imaizumi Asuka, 'Tōkyō ni okeru kōjō yō kikan to sono seizōgyōsha – 1889 kikan shurui torishirabe' (Factory boilers and their manufacturers: An investigation based on the 1889 boiler survey in Tōkyō), in *Gijutsu to bunmei*, vol. 21 no. 2, 2017, pp. 1–20.

from private businesses. A combination of these research findings and existing knowledge puts boilermakers in Japan in the 1880s through to the 1890s into the following three types.

3.1. *Type One: Government-run factories and the large private factories that came into being as a result of government disposals.*[10]

There was very significant progress in boiler-making in the second half of the nineteenth century in the field of naval shipbuilding that had been the catalyst for their introduction into Japan in the first place as they moved from the fire-tube to the water-tube stage. The Navy wanted to bring in the latest technology and, with this in mind, provided instruction by French naval engineers at its Yokosuka dockyard and also sent trainee engineers to study in France and England, as a result of which boilers were made there with the help of these foreign engineers and their fully-trained Japanese counterparts. Water-tube boilers relied on imports to start with but in 1903 the Navy decided to adopt the Miyabara Boiler, relatively easy to make, designed by Miyabara Jirō (1858–1918) who had studied marine engineering in England. This marked the start of domestic production.[11]

Five of the land-based boilers in Tōkyō were products of the Ministry of Public Work's Akabane Engineering Works which opened in 1875 and which also served as a practical training ground for students of the Imperial College of Engineering under the supervision of the foreigners, among them many Scotsmen, who had come to Japan to be on its teaching staff. In 1875, a boiler, which had a brick chimney and flue was installed there and was cladded with fire proof, white bricks. These were made by the Ministry of Public Works and were also supplied to the commercial sector via the Akabane Engineering Works.[12] In and

[10] These were the Mitsubishi Shipyard in Nagasaki, the Kawasaki Shipyard in Kōbe, the Ishikawajima Shipyard in Tōkyō and the Tōkyō Machinery Manufacturing Company. To these should be added the large factory set up by the Shibaura Manufacturing Works that prospered as a result of orders from the Navy at a time when torpedoes were being introduced Shibaura seisaku-sho (Tōkyō Shibaura denki KK) (ed.), *Shibaura seisaku-sho 65nen shi* (Sixty-five years of the Shibaura Manufacturing Works), 1940, pp. 16–17.

[11] Yamaguchi Susumu, 'Miyabara suikan shiki boiraa to sono seizō gijutsu ni tsuite' (The Miyabara water-tube boiler and its manufacture), in *Kagakushi kenkyū*, vol. 29, no. 174, 1990, pp. 74–82.

[12] 'Kōbushō kōka jō 8, Seisaku-ryō, Meiji 8 ichi gatsu kara roku gatsu made' (Ministry of Engineering's operating statement no. 8, Design Bureau, January–

around the Tōkyō of the time ordinary red bricks were a common sight for the public and with the purchase of boilers and fire bricks from the government factories, boilers with proper brick-built structures attached could be installed.

Progress continued and with Japan's first domestically-produced steam engine appearing in 1893 domestic production of machinery, including boilers, progressed to the point where the technology was approaching world standards. These were produced under the supervision of foreign engineers at government factories and large commercial enterprises that derived from government disposals, as well as by Japanese who had studied abroad or who were graduates of home universities.[13]

It was in 1879 that the first students in that field graduated from the mechanical engineering department at the College of Engineering and, among the 39 students who had graduated by 1885, nine had taken boilers as the subject of their theses. Of these, four took the 'locomotive boiler', one the 'marine boiler', and the remaining four 'land stationary boilers', as their subjects. The 1885 graduates among them were treating the 'steel boiler' that had lately become the common type.[14] In the early days of university education there was much interest in the design of boilers and this served as an incentive for domestic production. But this did not mean that imports ceased, especially as there was virtually no production possible of iron and steel plate or of iron pipe, all of which still had to be come from abroad.

3.2. *Type Two: Independent development of the commercial factory*

The only equipment needed for boiler production at the time was a roller for bending iron plates and, once plates had been bent, manual labour and tools for moving the plates around were used. The 1880s saw craftsmen who had learnt their trade in government factories or ones managed by foreigners starting up their own businesses, and there were also cases where people moved to commercial shipyards or machine plants to make boilers.[15]

June 1875) (preserved in the National Archives).

13 Suzuki Jun, *Meiji no kikai kōgyō – sono seisei to tenkai* (The Machine industry in the Meiji period – its origin and development), Tōkyō: Minerva shobō, 1996, pp. 109–111.

14 English text, kept in the Library of Tōkyō University's Engineering Building no. 2.

15 Similar movements of craftsmen occurred across a wide range of sectors in mechanical engineering. See Sawai Minoru, 'Kikai kōgyō' (Mechanical

Imaizumi points out that in Tōkyō in 1889, the two commercial factories whose products can be identified – three units came from one of them and two from the other – both started out by focusing on machinery for shipbuilding and marine use, and that at least four domestically-made boilers, including one of the five just mentioned, were marine boilers that had been installed on land. She also argues that the production of marine boilers formed the basis for land-based boiler manufacture at small and medium commercial enterprises.[16]

In the Ōsaka-Kōbe area too there was a flurry of such factories opening for business, an example of which was one specialising in boiler production, which was opened in Ōsaka in 1887 by Ōi Kenshirō, who had gained a wealth of experience at Nagasaki.

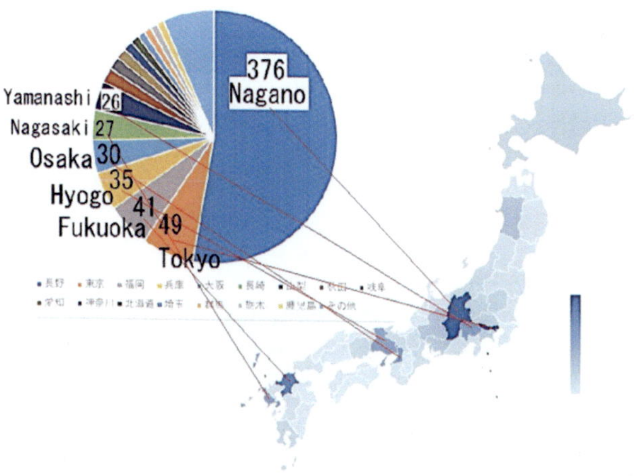

Fig. 3: Boiler-producing factories with ten or more employees in Hokkaidō and in the prefectures in 1902

a) Prefectures with factories specialising in boilers or manufacturing boilers among other products (dark blue in map)

b) Prefectures with factories manufacturing boilers among other products (mid-blue)

c) Prefectures without boiler-producing factories (light blue)

Source: Nōshōmushō shōkō-kyoku kōmu-ka (ed.), *Kōjō tsūran Meiji 35nen* (Factory Gazetteer Meiji 35 (1902)), Tōkyō, 1904.

engineering), in Nishikawa Shunsaku, Abe Takeshi (eds), *Nihon keizaishi 4, Sangyōka no jidai* (Japanese economic history 4, The age of industrialization) vol.1, Tōkyō: Iwanami shoten, 1990, pp. 214–253, esp. p. 217.

16 See Imaizumi, 'Tōkyō ni okeru kōjō yō kikan to sono seizōgyōsha', pp. 13–18.

Here as well there were instances of a parallel business in marine boilers.[17] Craftsmen moved around the country following the trail of demand for their services and by the end of 1902 fourteen of the country's forty-seven administrative areas were identified as having boiler-making factories employing more than ten artisans. These were confined to urban areas (Fig. 3).[18]

3.3. *Type Three: Locally produced boilers*

In 1886, immediately after Japan's industrial revolution had started, boilers were used in seventy-eight cases for railway steam locomotives, fifty-three on the government railways and twenty-five on private railways, in twenty-two of the Navy's capital ships and ten torpedo boats and small craft; in contrast, there were 460 steam ships in private hands.[19] As for the factory and mining sector, it had been assumed that there were about 100 boilers in use in the eleven factories and five mines run by the government, but in a study by the Ministry of Agriculture and Commerce, 332 were identified in the private sector including the seventy-eight that were not paired with a steam engine.[20] Ship-borne and land-based boilers vied with each other for application, with use in general concentrated in the private sector, though the Ministry of Agriculture and Commerce's study does not include Nagano prefecture due to their survey there being 'incomplete'. Instead, if one uses the study of silk mills by the Nagano prefectural Sericulture Industry's Union Management Office, 376 of them used steam, and three of those also used steam machinery.[21] On the basis of one per factory, taken together with twenty-six in neighbouring Yamanashi prefecture and twelve in Gifu, 373 units were exclusively for heating. A boiler used for heating does not have to with-

[17] See Suzuki, *Meiji no kikai kōgyō*, pp. 121–124, 133–134.

[18] Nōshōmushō shōkō-kyoku kōmu-ka (ed.), *Kōjō tsūran Meiji 35nen* (Factory Gazetteer Meiji 35 (1902)), Tōkyō 1904, pp. 211–219. There are also cases where boilers were made at factories belonging to coal mines and at factories employing less than nine workers.

[19] Naikaku tōkei-kyoku (ed.), *Nihon teikoku dai 7 tōkei nenkan* (The Japanese Empire's Statistics Annual 7), Tōkyō 1888, pp. 316, 326–327, 346–347, 745–748.

[20] Nōshōmushō (ed.), *Dai 2ji Nōshōmushō tōkei hyō* (Second agriculture and commerce table of statistics), Tōkyō 1887, pp. 402–408.

[21] Nagano-ken sanshigyō kumiai torishimari sho (ed.), *Nagano-ken sanshigyō kumiai torishimari-sho nenpō dai 2 kai* (Annual report of the central office of silk industry cooperatives in Nagano prefecture, 2nd Edition), 1888, seishi dai 7 hyō (Silk Manufacture Table 7).

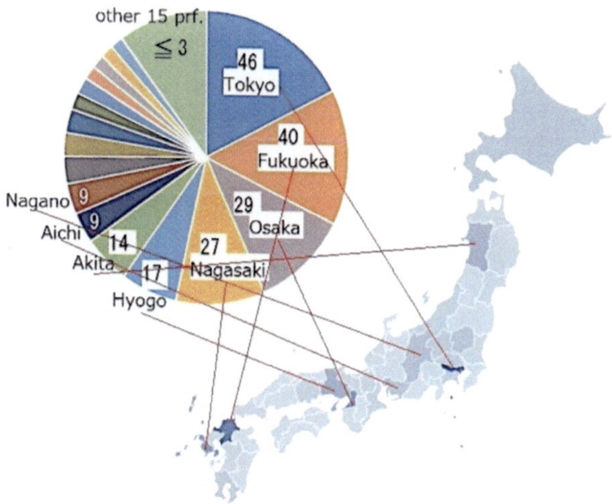

Fig. 4: Boilers used for motive power in the prefectures in 1886. As the colours show, the biggest concentration was in Tōkyō, followed by the prefectures Fukuoka, Ōsaka and Nagasaki.
Source: Nōshōmushō (ed.), *Dai niji Nōshōmushō tōkei hyō* (Second agriculture and commerce table of statistics), Tōkyō 1887

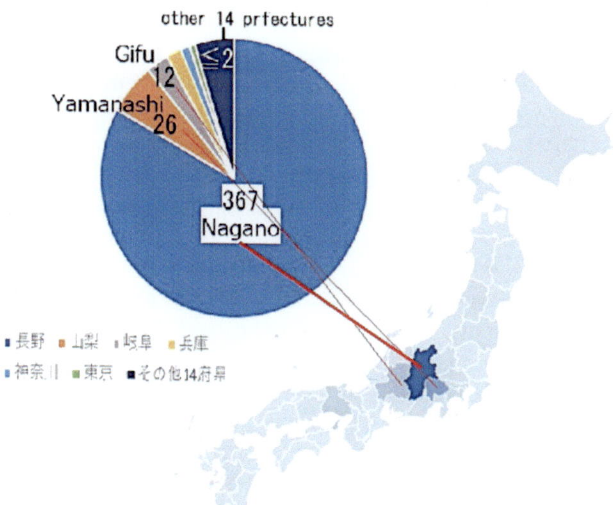

Fig. 5: Boilers not used for motive power in the prefectures in 1886. The dark colours show, that most of them were concentrated in the prefectures of Nagano, Gifu and Yamanashi in central Japan, followed by Hyōgo and Iwate.
Source: Nōshōmushō (ed.), *Dai niji Nōshōmushō tōkei hyō* (Second agriculture and commerce table of statistics), Tōkyō 1887, and Nagano-ken sanshigyō kumiai torishimarisho (ed.), *Nagano-ken sanshigyō kumiai torishimarisho nenpō dai 2 kai* (Annual Report, 2nd Edition), 1888.

stand high pressures and it can be of lighter construction than one driving a steam engine, and so this type developed independently. Distinguishing between boilers for power and non-power generation and comparing them in the prefectures where they were sited, Figs. 4 and 5 show that there are noticeable differences, with non-power generators used mainly in inland areas where there were no boiler-making factories employing more than ten workers, in which case their products were mainly supported in the early period by the existing local metal-working industry. Nagano led the way nationally in factory-based silk manufacture, which made this production flow particularly important.

4. SILK MILL BOILERS AND THEIR PRODUCTION

Nagano prefecture was unusually advanced compared to the rest of Japan in its factory-based industrialization of its silk-reeling industry. Silk reeling involves boiling the silkworm cocoons to loosen them and then withdrawing the fibres. In the process of drawing out the fibres the cocoons have to be kept at a constant temperature. Otherwise, once they have been heated to make them unravel, they will re-harden. Therefore, heating is needed for boiling the cocoons and for the process wherein women workers draw the fibres out of the cocoons, twist them together and reel them in as raw silk.

In 1805 France installed boilers in its silk factories. They fed heat to pans that each silk worker had next to them, thus making it easy to maintain a constant temperature for boiling the cocoons and reeling the silk. This method spread throughout France and in the 1870s it became the norm in Italy.[22] At the beginning of the 1870s Japan introduced European silk-reeling methods to improve the productivity and quality of its processes and this entailed two systems. The first was the Italian system, which was installed in 1870 under a Swiss engineer with experience gained in Italy, and this used a furnace for heating. In this system two silk reelers and one person who boils the cocoons sit facing each other across a brick-built structure that combines a furnace and a work surface. A fire is set under the pan that boils the cocoons, and the steam from it also heats the pans of the workers reeling the silk, as it makes its way to a brick-built chimney via a duct made for the purpose. The other system was the one introduced

22 Federico, Giovanni, *An Economic History of the Silk Industry, 1830–1930*, Cambridge: Cambridge University Press, 1997, pp. 104–113.

by a French engineer in 1872 to the Tomioka silk mill, which has since become a UNESCO heritage site. This system uses a boiler for heating. The Tomioka Mill itself used a steam engine for its motive power and had imported a fully-fledged boiler from France for this purpose.

How to involve the previous ruling class in industry was a point of national debate during this period. To achieve this, where the heads of the old fiefs or the provinces provided funds, French-style factories copied from the one at Tomioka were set up and full-scale boilers installed. However, as this was a time when boiler use had started to attain prominence in Italy too, and as French trading houses in Yokohama were sceptical of the chances for their boilers in Japan as late as 1875, the government was advised that it would be better to promote the Italian-style model.[23]

Fig. 6: A boiler used in Suwa in the early period, which is a traditional-pattern oven fitted with an iron lid (135cm x 90cm). Source: Nagano-ken Suwa-gun Hirano-mura yakuba (ed.), *Hirano sonshi* (A history of Hirano village) vol. 2, Okaya-shi, 1932.

[23] In a letter from Hecht, Lilienthal & Co dated 5 April 1875 to the Foreign Minister, in Nōshōmushō nōmu-kyoku ed., *Nōmu tenmatsu* (Agricultural records), vol. 3, 1955, p. 872.

In 1872 and 1873 Nagano prefecture saw factories built in one place after another with Italian-style boilers. In 1874, however, the former samurai family of Matsushiro, having sent a manager to the Tomioka silk mill to conduct a survey, founded the Rokkō-sha silk mill and had local coppersmiths make a boiler of copper plates for use there. Then, in 1875, farmers in Suwa who had examined this turned to the existing foundry industry to make a boiler in the style of the large iron cauldrons used for fermenting soya beans for making *miso* or cooking rice for large numbers of people, with a lid added (Fig. 6). This was a success and this type of boiler became popular.[24] Why was it, I wonder, that the prospects of foreign merchants were dashed? It could have been because of the considerable difficulty in making brick ovens in the Nagano prefecture at the time.

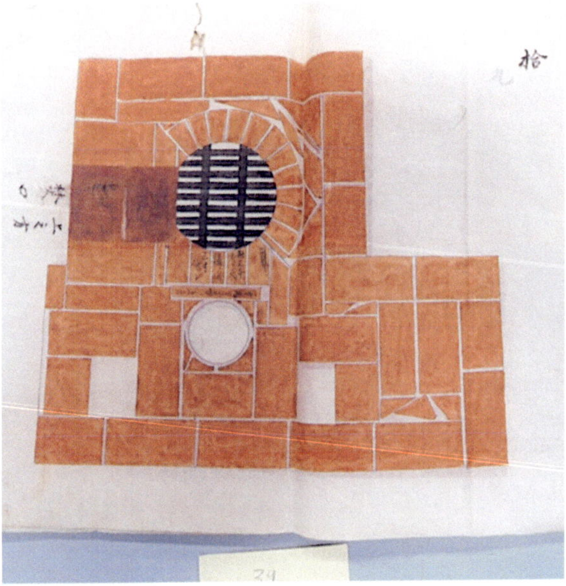

Fig. 7: The tenth in a series of 17 illustrations of the Italian-style furnace at the Ministry of Public Works' Silk Mill.
Source: Tōkyō Nōkō Daigaku Kagaku Hakubutsukan (Science Museum, Tokyo University of Agriculture & Industry)

24 Eguchi Zenji, Hidaka Yaoshichi (eds), *Shinano sanshigyō-shi* (A history of the silk industry in Shinano), Nagano: Dai Nihon sanshikai Shinano shikai 1937, pp. 940–941.

Not long ago a picture of the furnace in an Italian-style factory, which had been opened in Tōkyō by the Ministry of Public Works in 1873, was identified.[25] This is believed to be have been made by craftsmen sent from what was then Oda prefecture to set up a silk mill in Kasaoka in today's Okayama prefecture.[26] Fig. 7 shows the tenth of 17 individual drawings, on each of which the bricks of a cross section are painted. The gaps between the bricks are filled with crushed tiles. This is also the same where the chimney and the sunken smoke duct joining the furnace and the chimney are concerned, where the illustration shows the brickwork row by row. This kind of preparation was necessary for people with no experience of bricklaying. Oda prefecture sent its manager to factories in Tōkyō to make this plan drawing but this kind of exercise would have been difficult for a private company to undertake. To make the smoke duct function properly it would have been necessary to block the gaps between the bricks with mortar to prevent air getting in and to regularly repair any damage or warping; this too would have been difficult for an inexperience person. At Kasaoka bricks were brought in by sea from the area around Ōsaka, where they were already being fired, to build Western-style buildings, but for an inland prefecture like Nagano their transportation was a problem of a different order. When the Tomioka silk mill, which was nearer to Tōkyō, was being built, specialists were brought in to fire bricks near the construction site.[27]

In 1872 objects, which were probably bricks, were fired in Suwa in preparation for the arrival of an Italian-style boiler[28] while in 1874 the boiler at Matsushiro's Rokkō Company was not bricked in but coated with clay, a covering that was cracked open and replaced whenever the boiler was inactive.[29] Starting in

25 In the collection of the Tōkyō Nōkō Daigaku Kagaku Hakubutsukan (Science Museum, Tokyo University of Agriculture & Industry).

26 *2017nendo Seiji-keizaigaku keizaishi gakkai, shūki gakujutsu taikai, sōkai, hōkoku yōshi* (The 2017 summary report of the autumn congress cum general meeting of the Society for Political Economy and Economic History), 2017, pp. 6–7.

27 Tomioka seishijō shi hensan iinkai (ed.), *Tomioka seishijō-shi,* jō (A history of the Tomioka silk mill), vol. 1, Tomioka: Tomioka-shi kyōiku iinkai, 1977, pp. 26–29.

28 Nagano-ken Suwa-gun Hirano-mura yakuba (ed.), *Hirano sonshi* (A history of Hirano village), Okaya-shi, 1932, pp. 333–334.

29 Yokota Takeshi, 'Meiji 7nen 7gatsu yori 12gatsu made dai Nihon teikoku minkan jōki kikai no ganso roku kōsha sōritsu dai ichi nen no maki, seishigyō no ki' (An account of the first year of Rokkōsha's founding, the pioneer of steam machinery

1875 a type of boiler consisting of a large iron pot-like receptacle became common and opinion has it that it was fitted with a furnace built of stones and clay. There is an instance in Suwa in 1878 of a forge-master employing twenty people to make a boiler with a chimney 3.6 meter high, but this was built by a tile-maker and no bricks were used.[30] The purpose behind bundling a number of clay tubes together to make a chimney on top of the furnace was no doubt to improve efficiency and to stop smoke blowing back into the factory. Furnaces with fitted chimneys gained currency as the 'improved model' of the year, but it is not known what they looked like nor to what extent they spread at that particular time. The fact remains, however, that as long as brick-built furnaces were difficult to make, Japan continued to be influenced by Western models in its development of boilers with vessels and furnaces currently in use.

We know that numerous forge-masters in Yamanashi and Nagano prefectures made iron vessels to existing patterns, sometimes topped with iron lids or wooden buckets. There was also a type where metal flues were passed through a wooden container shaped like a Western-style barrel (Fig. 8). Established

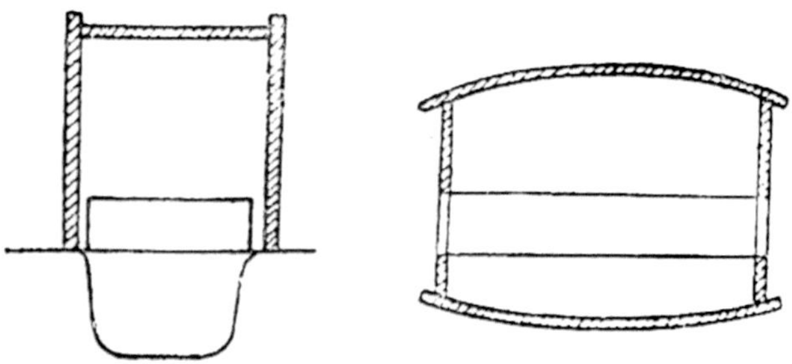

Fig. 8: A boiler consisting of a traditional-pattern oven capped with a tub, and one made from a wooden barrel with a flue inserted.
Source: Saitō Isamu, *Kikan shōtan* (Selected conversations on boilers), Shadan hōjin Nihon boiraa kyōkai, 1976.

in the private sector from July to December Meiji 7 (1874)), 1908, in Tomioka seishijō shi hensan iinkai (ed.), *Tomioka seishijō shi*, jō, 1977, pp. 875–902, esp. p. 884. See also Wada Eiko, *Tomioka Kōki* (The days after coming home from the Tomioka silk mill), Kokin-shoin,1931, p. 31.

30 Takayama Ryūzō, "Meiji 10nendai ni okeru seishi shihon no seisei to sonraku kōzō III" (Development of silk manufacture and a change of a village structure in the early Meiji era III), in *Mita gakkai zasshi*, vol. 62, no. 6, 1969, p. 113.

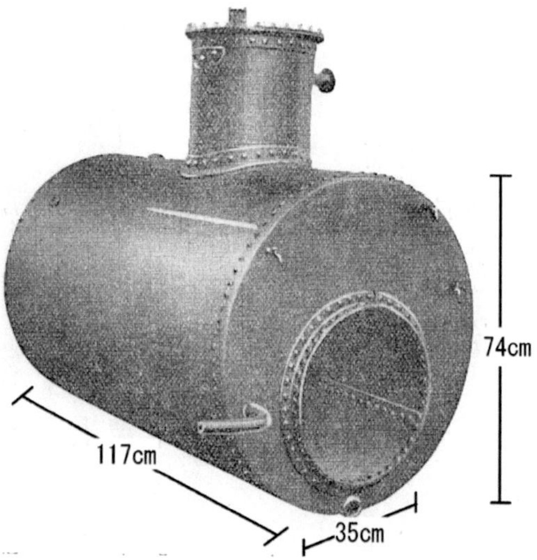

Fig. 9: A thin-plate boiler used in Suwa (117cm x 35cm x 74cm). Compared to what is shown in the House of Maruyama's boiler manufacturing records, the circumference matches its early-period thin-skinned boilers, but the length is somewhat shorter.
Source: Nagano-ken Suwa-gun Hirano-mura yakuba (ed.), *Hirano sonshi* (A history of Hirano village) vol. 2, Okaya-shi, 1932.

metal-working firms in both prefectures came up with a host of ideas and made boilers for use in factories with up to twenty silk-reelers.[31]

However, due to the size limits these boilers were subject to, they were superseded in the early 1880s by a boiler made of thin iron plates that resembled a Cornish boiler (Fig. 9) and then in the latter part of that decade by the multi-tube semi-continuous type (Fig.10). These were similar in construction to European and American boilers and made from materials imported from there. However, their working pressure was low so the steel plates from which they were made were thinner than in a fully-fledged boiler. A leading figure in the development of this type was Maruyama Yasorō (Yahei), a manufacturer of copper vessels

31 Suzuki, *Meiji no kikai kōgyō*, pp.145–147. Saitō Isamu, *Kikan shōtan* (Selected conversations on boilers), Shadan hōjin Nihon boiraa kyōkai, 1976, pp. 51–54.

図6-2 多管半通式汽罐

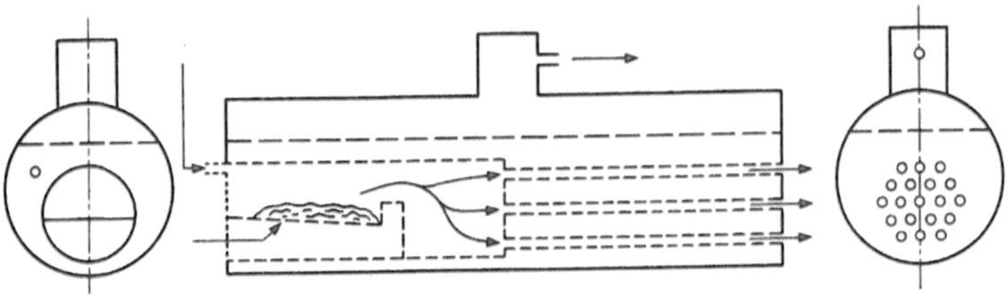

Fig. 10a: Boiler for cocoon heating: compound cornish multi-tube semi-continuous boiler (drawing)

Fig. 10b: compound cornish multi-tube semi-continuous boiler (as in Fig. 10a; Photo) Source: Suzuki Jun, *Meiji no kikai kōgyō – sono seisei to tenkai* (The machine industry in the Meiji period – Its origin and development), Tōkyō: Minerva shobō, 1996.

in Matsumoto since the Edo period, who also made the 'large-pot' type of boiler.[32]

This thin-skinned boiler was developed jointly in 1879 by Maruyama and Futaki Ryūzō, a silk industrialist on the outskirts of Matsumoto, after teething troubles had been resolved. In 1878

32 For information about Maruyama's operations see Suzuki, *Meiji no kikai kōgyō*, pp. 151–158.

Futaki Tangyū, who was a close relation of Ryūzō's and who lived in the neighbourhood, brought in a boiler made at the naval shipyard at Yokosuka. It is very likely that this was a Cornish boiler, but even if this was not the case, it was probably developed by referring to boilers, based on Western patterns. This could be used to accommodate as many as six silk-reelers. Starting in 1886 or thereabouts a larger version of the multi-tube boiler was made; this was a simplified version of the replacement boiler used in the warship Chiyoda, which was built in 1877 at the Ishikawajima naval shipyard (Fig. 11), and it was probably made by referring either to an actual example of this kind of marine boiler or an illustration of it.

On the subject of these thin-skinned boilers, we still have the ledgers recording the cost for a twenty-person reeler in 1881 and a forty-person example in 1888 from when the factories they

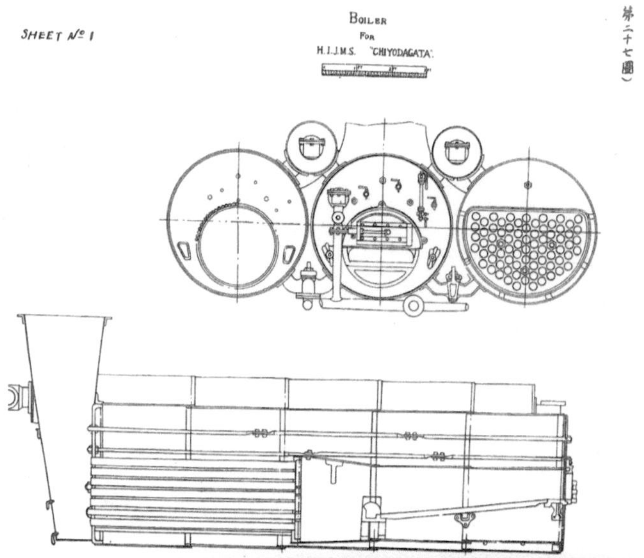

Fig. 11: Boiler in the warship *Chiyoda.* This compound cornish and multitubular boiler was made in 1877 by the naval dockyard in Ishikawajima, Tōkyō, for the Japanese warship *Chiyoda* to replace the older boilers, made in Saga in 1863.
Source: Kaigun kyōiku-kyoku, *Teikoku kaigun kikanshi* (A History of the imperial navy's machinery), 1937 and subsequent years.

were put in were built. They both have chimneys but there are no signs that bricks were used in their construction.[33] It is also possible that the brick-built flues on the outside of the boiler were omitted as the multi-tube non-continuous model's reheat function via flues from its interior in the same way as maritime or steam engine boilers was a highly-regarded feature. Finding illustrations or physical remains to determine how these boilers were installed has to wait for future research, but in Suwa chimneys made of iron plates were the norm until a large one built of bricks appeared in 1907,[34] which makes it a distinct possibility that in the mid-nineteenth century these boilers were installed without any, or just the bare minimum, association with brick-laying.

This then is the third type of boiler manufacturer: one who developed boilers to suit local conditions by adopting thin iron plates rather than bricks due to them being relatively easy to transport, and by introducing new materials and information based on current expertise. This is the point at which Kimura Daisuke makes an appearance in the manufacture of the multi-tube boiler. In 1886 he opens a business in Suwa, which was the centre of demand for what was the 'final manifestation' of that model in the area. Kimura had gained his experience in the army's arsenal in Tōkyō, which can be described as being of the second type, but in 1880 he had been sent by the forge-master Yamada Shōzaemon in Kōfu in Yamanashi prefecture to install a Cornish boiler in Suwa. Before embarking on boiler-making Yamada had a traditional metal working business that sent products to silk mills in Yamanashi prefecture, a situation not unlike Maruyama's. It was also possible to recruit craftsmen with a grasp of the new methods, to compensate for the old metal-working businesses. In the 1890s, alongside Maruyama and Kimura, Suwa's Kawanishi Torakichi, who had a background like Yamada's and experience as a blacksmith making agricultural implements, started making multi-tube semi-continuous boilers so that factories of both Type 2 and Type 3 were supplying boilers for use in silk mills.[35]

33 Ueda-shi no sakan gyōsha, 'Seishi kikai shinchiku sho jihikae' (Factors behind building new raw silk machinery) (property of the Saitō family at Kakeyu-onsen). Uehara Kazujirō of Kita-Azumi-gun, Ikeda-chō, 'Kikai fushin sho jihikae' (Tenets of Machinery Building) (property of the Uehara family).

34 Nagano-ken Suwa-gun Hirano-mura yakuba, *Hirano sonshi*, p. 346.

35 Suzuki, *Meiji no kikai kōgyō*, pp. 150, 155–156.

5. BOILER INSPECTION AND ITS SUPPORT STRUCTURE

Imaizumi pointed to the fact, that in 1889 when there were as yet no rules in place for inspection of land-based boilers some manufacturers made use of the publicity concerning the merits of inspection of marine boilers, and users of boilers were complying with the rules for their use set out by the Maritime Bureau at the Ministry of Communications. One could argue in this instance that the policy to ensure the safety of marine engines benefited the spread of land-based boilers.[36] As explained, many steamships were imported from the 1860s but the loss of seventy-three lives in the sinking of the *City of Edo*, a steamship owned by an American and which ran between Tōkyō and Yokohama, after an explosion from a fractured boiler, was a graphic reminder of the inherent dangers of boilers.[37] Then in 1880, following an increase in the number of commercial and private steamships mainly in western Japan,[38] the Ministry of Home Affairs ordered every prefecture to draw up and enforce a set of rules that reflected their particular situations, citing as an example steamship inspection regulations that included regular annual boiler checks. In 1881, in response to this Ōsaka prefecture carried out regular checks of steamers in the fourteen prefectures of western Japan and brought in a system for restricting boiler pressures. Steamships in this part of Japan frequently put into Ōsaka and so it was only logical that they should be inspected there. Similar rules were formulated in Tōkyō in the same year and in Niigata in 1882.[39] It was in 1884 that a complete shipping inspection system that included larger steamships was instituted by the Ministry of Home Affairs, preceded by the one covering the smaller steamships frequently used by medium and small operators whose technical level was low.[40]

---

36 See Imaizumi, 'Tōkyō ni okeru kōjō yō kikan to sono seizōgyōsha', pp.15–18.

37 'Tōkyō Tsukiji kaigan ni oite Amerika koku shō kisen 'Shichii ofu Eto'-gō kikan haretsu norikumi-in shi shō ikken' (Crew killed and wounded on board the American steamer 'City of Edo' off Tōkyō's Tsukiji, vol. 1 (B-3-6-3-9_001) JACAR Ref. B11092372400.

38 Yamazaki Zenkei, *Setouchi kindai kaiunsō sōshi* (The origins of marine transportation in the Setonaikai), Sōfū-sha, 2006, pp. 13–55.

39 Naikaku kanpō-kyoku (ed.), *Meiji 13nen hōrei zensho*, (Compilation of laws for 1880), 1890, pp. 945–947. Furuya Shūsaku, *Ruiju Ōsaka-fu futatsu zensho dai jūkan* (Collected regulations of Ōsaka prefecture vol. 10), Ryūunsha, Ōsaka, 1885, pp. 122–135. Suhara Tetsuji, *Tōkyō keishi honsho futatsu zensho Meiji 14nen* (Tōkyō Police Headquarters complex of regulations 1881), Tōkyō, 1882, pp. 80–94. Nagai Hidetarō, *Kaisei Niigata-ken sho kisoku binran* (Compilation of amendments to regulations in Niigata prefecture), Niigata: Bunrin-dō, 1884, pp. 20–32.

40 Teishinshō (ed.), *Kaiji* (Maritime matters), 1914, p. 7.

Table 1: Boiler failures as recorded in the *Kanpō* (Government Gazette) of various years

Table 1 Boiler failures as recorded in the Government Gazette					
Date	**Kanpō No.**	**Prefecture**	**Use / Type**	**Killed**	**Injured**
8/3/1889	1707 / 1712 / 1725	Ishikawa, Kanazawa District	Silk mill	9	8
25/3/1889	1724	Kyoto, Umezu Village	Paper mill / steam furnace made of forged iron	1	More than 4
11/9/1889	1864	Tokyo, Ishikawajima Shipyard	Steamship	7	5, 6
21/5/1892	2674	Nagano, Suwa District, Hirano Village	Silk mill	0	1
27/7/1892	2732	Nagano, Kami-Ina District, Miyata Village	Silk mill / made of wood	2	5
14/8/1892	2745	Nagano, Suwa District, Shimo Suwa Village	Silk mill	1	1
6/12/1892	2838	Hyogo, Saji Village	Silk mill	2	1
26/12/1892	2854	Nagano, Suwa District, Hirano Village	Silk mill	4	11
29/8/1893	3052 / 3491	Kagawa, Kagawa District. Miyawaki Village	Paper mill	0	5
8/7/1894	3313 / 3491	Fukuoka, Inaba District, Inaba Village	Coal mine	1	15
3/11/1894	3491	Nagano, Suwa District	Silk mill	0	0
3/11/1894	3491	Nagano, Ina District	Silk mill	0	0
31/7/1895	3635	Saga, Higashi-Matsuura District, Kita-Namita Village	Coal mine drainage	0	6
11/11/1895	3718	Osaka, Osaka City	Paper mill / steam furnace	1	1
6/12/1895	3746	Offshore Shimane	Steamship	3	2
30/9/1897	4285	Nagano, Kami-Ina District, Nakazawa Village	Silk mill	0	1

Inspections of land-based boilers became an issue as a result of a boiler bursting at a spinning mill at Kanazawa, Ishikawa prefecture on 8 March 1889, which killed nine people including women workers who were warming themselves in the boiler room before their shift. The boiler had been bought in Tōkyō in 1886 and was only used for heating. However, it used the iron plating of an older boiler and was smaller than the original, which resulted in problems with its construction and riveting.[41] Table 1 summarises information from *Kanpō* (Government Gazette) on instances of boilers bursting in the eight years following that accident. There were also frequent newspaper reports urging caution by users of boilers.[42] Active coverage in the *Gazette* was as a direct result of a series of major accidents in 1889, but was also because, as education in industrial matters advanced, the Government was able to make an accurate assessment of their circumstances and could take suitable action.

In 1886 the College of Engineering merged with Tōkyō University to become the Imperial University. Of the fourteen graduates the Imperial University's Mechanical Engineering Department had produced by 1890, only Takai Suketarō wrote his thesis on the subject of boilers. Theses written earlier at the College of Engineering were on the history of boilers or practical aspects of design, such as calculations of boiler strength but Takai's *An essay on the economical and safe use of steam* differed. As the rapid spread of boilers had left Japan with a body of inexperienced users, his thesis argued the case for a selection based on economic reasons, considering a boiler's fuel and design, and pointed out the importance of regular inspections by the government. It also brought up the subject of boiler ruptures in Europe and the United States and their systems of boiler insurance.[43] On graduation Takai became an engineer in the Ministry of Agriculture and Commerce and investigated the boiler failure in Kanazawa the following year. His inspection report was published in *Kanpō* (Government Gazette) no. 1725.

Kyōto prefecture, which only had nine boilers at the time, was the first administrative area to formalise a regular annual system of boiler inspections when it introduced its regulations in January 1887 and settled on April as the month for them to be made.[44] Tanabe Kitarō and other graduates of the College of Engineering who worked at the prefectural government on the construction of the Lake Biwa Spillway almost certainly had a hand in having the regulations drawn up and the inspection regime put in place, but even so there was a boiler failure in Kyōto two years

41 *Kanpō* (Government Gazette) no. 1725, 1889, pp. 31–32.

42 *Yomiuri shinbun*, e.g. 31 March 1889, 3 October 1889 etc.

43 In English. Kept in the library of Tōkyō University's Engineering Building no. 2.

44 Kyōto-fu shōmu-ka (ed.), *Kyōto-fu fureitatsu yōyaku* (Summary of Kyōto prefecture's Regulations), 8th Edition, 1888, pp. 1–3.

later, as can be seen from Table 1. Whether or not it was because this accident occurred in the month preceding its inspection, the National Police Agency, which was responsible for control in Tōkyō and set its own rules two months later, made six-monthly checks mandatory.[45] The first such inspections were made of 108 boilers in June 1889 and were carried out by Professor Taniguchi Naosuke (who graduated from Glasgow University in 1878) of the Imperial University's Engineering Faculty and Yoshida Tomoyoshi (who graduated from the Mechanical Engineering Faculty of Tōkyō University's Physics Department in 1881) who served as an engineer for Kanebō (Kanegafuchi Boseki KK, a large spinning company) and a lecturer at the College of Engineering.[46] In Ōsaka prefecture, where the inspection system was set up the following year, Satachi Jirō carried out these inspections on a regular basis thereafter. Satachi had graduated ten years earlier from the College of Engineering and had gained experience at the Ministry of Public Works' Nagasaki Shipyard.[47] Inspections of land-based boilers were conducted by university-trained engineers with experience of actual working conditions. They were first carried out in large cities where boilers were concentrated and where any accident would impact more than immediate operational staff. In provincial regions there were problems in two areas: (1) enlisting inspectors and (2) the standard of the boilers themselves.

Throughout 1889 Takai surveyed conditions in Gunma and Nagano prefectures and lectured to silk millers at Suwa Police Station. He pointed out that there were no boiler manufacturers in Suwa whose products were based on scientific principles and that their design and the materials used were both inadequate. Moreover, the people who handled them were not sufficiently knowledgeable. His opinion was that a thorough inspection would not find a single boiler in the prefecture that would pass, a situation in which the industry would obviously not prosper. He advised that priority be placed on maintenance and proper handling, and predicted that, without this, boiler failures would continue to occur.[48]

---

45 Koishikawa keisatsu-sho (ed.), *Meiji 22nen 5gatsu keisatsu yōmu* (May 1889, Police priority activities), 1889, pp. 1–4.

46 *Kanpō*, no. 1825, 30 July 1889, pp. 7–8.

47 Satachi Jirō, 'Ōsaka-fu ka ni okeru jōkikan ni tsuite' (A Discourse on Steam Boilers in Ōsaka prefecture), in *Kikaigaku kaishi*, vol. 6, no. 9, 1903, pp. 21–37, esp. p. 23.

48 *Shinano Mainichi Shinbun*, 22, 23 January 1890.

Boiler failures had occurred in Nagano prefecture in 1880 claiming injuries and lives. Therefore, in 1881 seasoned manufacturers in Suwa District were asked to produce boilers to which they were advised to fit water-level gauges, pressure gauges and safety valves. In 1886 Nagano put regulations in place and prohibited the use of boilers that did not have these three safety features. These regulations also applied to inspections by the prefecture itself, but stated that if a guarantee was issued by a manufacturer who had been approved by the prefecture, an inspection was not necessary.[49] Given the results to date, this would no doubt have meant that security was assured, but the scheduled date of implementation was postponed from January 1888 to April that year.[50] Nevertheless, while Takai in Suwa was still criticising the use of these three safety devices as inappropriate, their actual fitting to boilers went ahead. As can be seen from Fig. 12, in 1888 the Nagano prefectural police identified sites where boilers were installed as one of their target areas for surveillance.

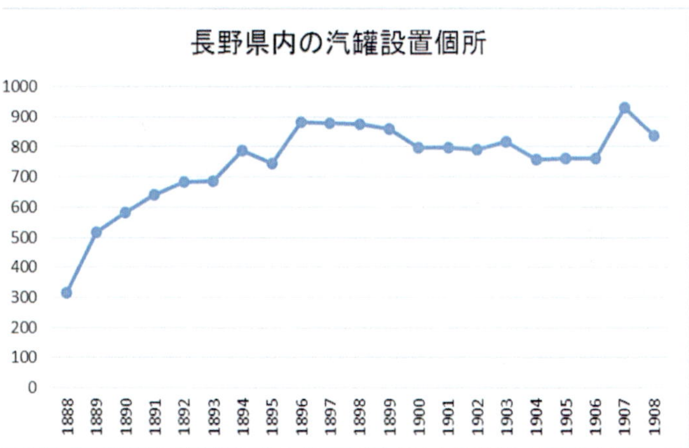

Fig. 12: Number of sites in Nagano prefecture with boilers installed.
Source: Yearly totals from Nagano-ken (ed.), *Nagano-ken tōkeisho* (Statistics of Nagano prefecture), Nagano, various years.

49 Suzuki, *Meiji no kikai kōgyō*, pp. 158–159.

50 Nagano-ken chiji kanbō, hoan-ka (ed.), *Nagano-ken genkō reitatsu ruisan, gekan* (A Compilation of the Current Legislation of Nagano prefecture vol. 2), Kashiyo Printery, 1899, p. 49.

In addition, taking a different approach from other prefectures and major administrative areas, the *Kanpō* (Government Gazette) included also accidents that did not involve personal injuries. The determination to bring boiler accidents under control is obvious. The entry for July 1892 in Table 1 relates to an accident involving a trial of a wooden boiler that provided power for four silk-reelers, and demonstrates the dangers inherent in attempting to come up with a simplified boiler for use in small-scale factories.

In a speech in Suwa Takai related how the female workers from other silk mills left hurriedly for home after the disaster at Kanazawa and that their male relatives did not make them go back to work. This, he explained, forced all the mills in Kanazawa to close down. He warned that the effects of a boiler failure were not restricted to that one factory alone but extended to the area's industry as a whole. In 1892 there was a spate of boiler failures as he had anticipated, and this could only have heightened silk industrialists' awareness of the danger. Consequently, as can be seen from Fig. 12, from the middle of the 1890s the number of boilers in Nagano prefecture ceased to increase. The number of mills fell due to competition while the scale of operations broadened in each factory. Linked to this broadening of scale were larger boilers with increased pressures, and an increase in the number of factories using steam, rather than water, as a source of power. Japan was therefore nearly at the stage where boilers were assumed to conform to European and American standards. Attitudes changed too, with users no longer relying on a manufacturer's experience for the safety of a boiler, but accepting the logic of an engineer's inspection as their assurance of security. Therefore, in 1901 Nagano prefecture finally adopted the regular boiler inspection system used by other administrative areas.[51]

In 1891 only the three major conurbations, Tōkyō, Ōsaka and Kyōto, carried out boiler inspections but, as Fig. 13 shows, the system then spread to other parts of the country so that by 1902 forty-three of the forty-six prefectures had it in place.[52] Engineers educated in industrial high schools played a major role

[51] Ozaki Shōichi, *Kōjō-hō kankei hōki kaisetsu* (An explanation of the legislation relating to Factory Law), Kōjō-hō kaisetsu shuppan-bu, 1918, p. 69–73.

[52] Nōshōmushō shōkō-kyoku (ed.), *Kōjō chōsa yōryō* (Summary of Factory Inspections), 1904, p. 37.

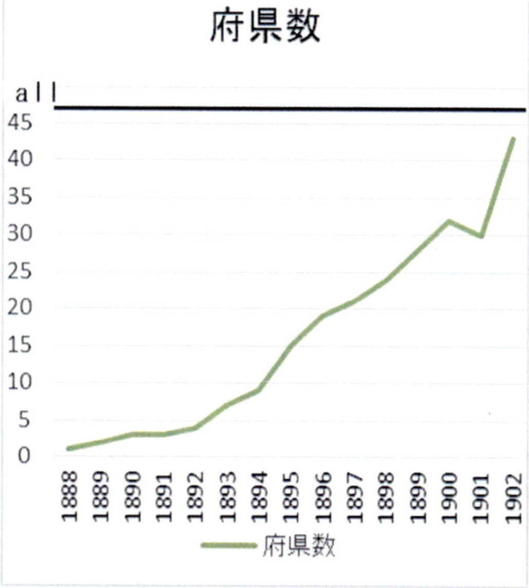

Fig. 13: Number of prefectures carrying out regular boiler inspections (all = total number of prefectures).
Source: Nōshōmushō Shōkō-kyoku (ed.), *Kōjō chōsa yōryō* (Summary of factory inspections) 1904.

as inspectors outside of the large cities and, by the end of 1906, eleven graduates from the Mechanical Department of Tōkyō's Industrial High School worked for prefectural governments or for the police.[53] In Nagano, for instance, Sugitani Shirō who left this high school in 1894 was employed as an engineer when these inspections were introduced.[54] In prefectures that did not have many boilers engineers were employed on a temporary basis for inspections.[55] Despite this, inspections were more lenient in the provinces than in Tōkyō or Ōsaka and the flow of second-hand boilers from the cities to outlying areas continued.[56]

53 Tōkyō kōtō kōgyō gakkō (ed.), *Tōkyō kōtō kōgyō gakkō ichiran* (The Tōkyō Higher Industrial College, review for 1906–1907), Tōkyō, 1906, pp. 84–98.

54 Insatsu-kyoku (ed.), *Shokuin roku, Meiji 13 nen, otsu*, (Record of employees, 1902, Part B), 1902, p. 243.

55 'Kiki kikan kensa enki' (Prolonged Inspection of Machinery and Boilers), in *Asahi shinbun*, 1903, 20 August 1903, front page, example from Ibaraki.

56 Haga Sōjirō, 'Rikujō kikan no anzen en fūsa ni tsuite', (Blockage of safety valves in land-based boilers), in *Keisatsu kyōkai zasshi*, no. 79. 1906, p. 2.

From around the beginning of the twentieth century illustrations such as the one in Fig. 2, which shows a boiler barrel and the brick-built structure surrounding it, began to appear in introductory works on the silk-reeling and distilling industries.[57] Until this time, this kind of tacit knowledge had been passed from craftsman to craftsman. In 1902 there were brick manufactories with more than ten workers in twenty-six of Japan's prefectures.[58] In addition, according to prefectural statistics there were four prefectures where bricks were made.

There was a close relationship between the craft of brick making and Japan's traditional kiln-based products such as tiles, and the technique of brick-making spread throughout the country faster than western-style boiler manufacture. This was why the development of boilers without this brick cladding was a trend that was confined to Nagano and the surrounding area where the booming silk industry led to the rapid development of factories.

6. CONCLUSION

We tend to assume that, between the opening up of the country and the advent of its industrial revolution, the modernization of Japan's industry was due to the latest Western technology being introduced and then used to promote a series of innovative ideas based on existing home-grown methods. Before then, however, the West and Japan had different cultures, which meant that the transfer and spread of Western pre-industrial-revolutionary artisan workmanship and skills such as brick making continued concurrently with this process and affected it accordingly. Not content with simply introducing new methods and improving them, engineers also concerned themselves with ensuring their safety. Engineers made this their first priority in the case of small steamships and conducted rigorous inspections, while in silk mills they investigated the circumstances of accidents caused by boiler failures and noted the changes boilers underwent. With the insight into the current state of industry and technology thus acquired, engineers were also enabled to reconcile the need

---

57 Yamamoto Takezō, *Nihon seishi hō* (Silk reeling in Japan), Tōkyō: Bunmei-dō, 1909, pp.140–141. Suzue Kintarō, *Jōzō oyobi setsubi* (Distillation and its equipment), Shōyu sekaisha, Tōkyō 1909, pp. 17–22.

58 Nōshōmushō shōkō-kyoku kōmu-ka (ed.), *Kōjō tsūran 1902*, pp. 225–245.

for improvements in safety standards with progress in industrial performance.

REFERENCES:

'Kōbushō kō-ka jō 8, Seisaku-ryō, Meiji 8, ichigatsu kara rokugatsu made' (Ministry of Public Works' operating statement no. 8, Design Department, January–June 1875) (preserved in the National Archives).

'Kiki kikan kensa enki' (Prolonged inspection of machinery and boilers), in *Asahi shinbun*, 20 August 1903, front page, example from Ibaraki.

'Tōkyō Tsukiji kaigan ni oite Amerika koku shō kisen 'Shichii ofu Eto'gō kikan haretsu norikumi-in shi shō ikken' (Crew killed and wounded on board the American steamer *City of Edo* off Tōkyō's Tsukiji, vol. 1 (B-3-6-3-9_001) JACAR Ref. B11092372400.

2017nendo Seiji-keizai-gaku keizai-shi gakkai, shūki gakujutsu taikai, sōkai, hōkoku yōshi (The 2017 summary report of the autumn congress cum general meeting of the political economy and economic history society), 2017.

Eguchi Zenji, Hidaka Yaoshichi (eds), *Shinano sanshigyō-shi* (A History of the silk industry in Shinano), Nagano: Dai Nihon sanshikai Shinano shikai, 1937.

Federico, Giovanni, *An Economic History of the Silk Industry, 1830–1930*, Cambridge: Cambridge University Press, 1997.

Furuya Shūsaku, *Ruiju Ōsaka-fu futatsu zensho dai jūkan* (Collected regulations of Ōsaka prefecture, vol. 10), Ryūun-sha, Ōsaka, 1885.

Haga Sōjirō, 'Rikujō kikan no anzen-ben fūsa ni tsuite', (Blockage of safety valves in land-based boilers), in *Keisatsu kyōkai zasshi*, no. 79, 1906, pp. 1–10.

Imaizumi Asuka, 'Tōkyō ni okeru kōjō yō kikan to sono seizōgyō sha – 1889 kikan shurui torishirabe' (Factory boilers and their manufacturers: An investigation based on the 1889 boiler survey in Tōkyō), in *Gijutsu to bunmei*, vol. 21 no. 2, 2017, pp. 1–20.

Insatsu-kyoku (ed.), *Shokuin roku, Meiji sanjūgonen, otsu* (Record of employees, 1902, Part B), 1902.

Ishiguro Nobuyoshi, *Sanpō tokai hyōteki* (Reference points for sea passages on arithmetical principles), ed. by Suharaya Mohei et al, 1836.

Kaigun kyōiku-kyoku, *Teikoku kaigun kikanshi* (A history of the imperial navy's machinery), 1937

Kanpō (Government Gazette), nos. 1725 and 1825, 1889.

Kattendyke [Willem Huyssen van Kattendijke], *Nagasaki kaigun denshū sho no hibi* (The story of the Nagasaki Naval Training Centre), translated by Mizuta Nobuyoshi, Tōkyō: Heibonsha, 1964.

Koishikawa keisatsu-sho (ed.), *Meiji nijūninen gogatsu keisatsu yōmu* (May 1889 police priority activities), 1889.

Kōshaku Shimazu-ke hensansho (ed.), *Sappan kaigunshi* (A history of the Satsuma navy) vol. 1, 1928.

Kusumoto Juichi, *Nagasaki seitetsusho* (The Nagasaki Ironworks), Tōkyō: Chūō kōron-sha, (Chūkō shinsho), 1992.

Kyōto-fu shōmu-ka (ed.), *Kyōto-fu fureitatsu yōyaku* (Summary of Kyōto prefecture's regulations), 8th Edition, 1888.

Nagai Hidetarō, *Kaisei Niigata-ken sho kisoku binran* (Compilation of amendments to regulations in Niigata prefecture), Niigata: Bunrin-dō, 1884.

Nagano-ken (ed.), *Nagano-ken tōkeisho* (Statistics of Nagano prefecture), Nagano

Nagano-ken chiji kanbō (ed.), *Nagano-ken genkō reitatsu ruisan,* gekan (A compilation of the current legislation of Nagano prefecture vol. 2), Nagano: Kashiyo Printery, 1899.

Nagano-ken sanshigyō kumiai torishimarisho (ed.), *Nagano-ken sanshigyō kumiai torishimarisho nenpō dai nikai* (Annual report of the central office of silk-industry cooperatives in Nagano prefecture, 2nd Edition), 1888, seishi dai nana hyō (Silk Manufacture Table 7).

Nagano-ken Suwa-gun Hirano-mura yakuba (ed.), *Hirano sonshi* (A history of Hirano village), Okaya-shi, 1932.

Naikaku kanpō-kyoku (ed.), *Meiji jūsannen hōrei zensho* (Compilation of laws for 1880), 1890.

Naikaku tōkei-kyoku (ed.), *Nihon teikoku dai 7 tōkei nenkan* (The Japanese Empire's Statistics Annual 7), Tōkyō 1888.

Nakanishi Hiroshi, *Nihon kindaika no kiso katei* (The fundamentals of Japanese modernization), vol. 1, Tōkyō: Tōkyō University Publishing Association, 1982.

Nōshōmushō (ed.), *Dai niji Nōshōmushō tōkei hyō* (Second agriculture and commerce table of statistics), Tōkyō 1887.

Nōshōmushō nōmu-kyoku (ed.), *Nōmu tenmatsu* (Agricultural records), vol. 3, 1955.

Nōshōmushō shōkō-kyoku (ed.), *Kōjō chōsa yōryō* (Parameters of factory survey), 1904.

Nōshōmushō shōkō-kyoku kōmu-ka (ed.), *Kōjō tsūran Meiji sanjūgonen* (Factory Gazetteer Meiji 35 (1902)), Tōkyō, 1904.

Ozaki Shōichi, *Kōjō-hō kankei hōki kaisetsu* (An Explanation of the legislation relating to Factory Law), Kōjō-hō kaisetsu shuppan-bu, 1918, p. 69–73.

Saitō Isamu, *Kikan shōtan* (Selected conversations on boilers), Shadan hōjin Nihon boira kyōkai, 1976.

Satachi Jirō, 'Ōsaka-fu ka ni okeru jōkikan ni tsuite' (A discourse on steam boilers in Ōsaka prefecture), in *Kikaigaku kaishi*, vol. 6, no. 9, 1903, pp. 21–37.

Sawai Minoru, 'Kikai kōgyō' (Mechanical Engineering), in Nishikawa Shunsaku, Abe Takeshi (eds), *Nihon keizaishi 4, Sangyōka no jidai*

(Japanese economic History 4, The age of industrialization) vol. 1, Tōkyō: Iwanami shoten, 1990, pp. 214–253.

Shibaura seisakusho [Tōkyō Shibaura denki KK] (ed.), *Shibaura seisakusho rokujūgonen shi* (Sixty-five years of the Shibaura Manufacturing Works), 1940.

Shinano Mainichi Shinbun, 1890.

Sugiyama Kenjirō, 'Nairin shiki jōkisen Sentō-maru ni tsuite' (The *Sentō -maru* screw steamer), in *Chiba shōdai ronsō*, vol. 40, no. 3, 2002. pp. 41–77.

Suhara Tetsuji, *Tōkyō keishi honsho futatsu zensho Meiji 14nen* (Tōkyō police headquarters complex of regulations 1881), Tōkyō, 1882.

Suzue Kintarō, *Jōzō oyobi setsubi* (Distillation and its equipment)', Tōkyō: Shōyu sekai–sha, Tōkyō 1909.

Suzuki Jun, *Meiji no kikai kōgyō – sono seisei to tenkai* (The machine industry in the Meiji period – its origin and development), Tōkyō: Minerva shobō, 1996.

Takayama Ryūzō, 'Meiji jūnendai ni okeru seishi shihon no seisei to sonraku kōzō III' (Development of silk manufacture and a change of village structure in the early Meiji era III), in *Mita gakkai zasshi*, vol. 62, no. 6, 1969, pp. 111–125.

Teishinshō (ed.), *Kaiji* (Maritime matters), 1914.

Tōkyō kōtō kōgyō gakkō (ed.), *Tōkyō kōtō kōgyō gakkō ichiran Meiji 39–40* (The Tōkyō Higher Industrial College review for 1906–1907), Tōkyō,1906.

Tomioka seishijō shi hensan iinkai (ed.), *Tomioka seishijō-shi* (A History of the Tomioka Silk Mill), vol. 1, Tomioka: Tomioka-shi kyōiku iinkai, 1977.

Ueda-shi no sakan gyōsha, 'Seishi kikai shinchiku sho jihikae' (Factors behind building new raw silk machinery) (property of the Saitō family at Kakeyu-onsen).

Uehara Kazujirō of Kita-Azumi-gun, Ikeda-chō, 'Kikai fushin sho jihikae' (Tenets of Machinery Building) (property of the Uehara family).

Wada Eiko, *Tomioka Kōki* (The Days After Coming Home from the Tomioka Silk Mill), Kokin-shoin,1931.

Yamaguchi Susumu, 'Miyabara suikan shiki boiraa to sono seizō gijutsu ni tsuite' (The Miyabara water-tube boiler and its manufacture), in *Kagakushi kenkyū*, vol. 29, no. 174, 1990, pp. 74–82.

Yamamoto Takezō, *Nihon seishi hō* (Silk reeling in Japan), Tōkyō: Bunmei-dō, 1909.

Yamazaki Zenkei, *Setouchi kindai kaiunsō sōshi* (The origins of marine transportation in the Setonaikai), Sōfū-sha, 2006.

Yokosuka kaigun kōshō (ed.), *Yokosuka kaigun senshō shi* (The history of the Yokosuka naval yard), vol. 1, 1915.

Yokota Takeshi, 'Meiji nananen shichigatsu yori jūnigatsu made dai Nihon teikoku minkan jōki kikai no ganso roku kōsha sōritsu dai ichi

nen no maki, seishigyō no ki' (An account of the first Year of Rokkō sha's founding, the pioneer of steam machinery in the private sector from July to December Meiji 7 (1874), 1908, in Tomioka seishijō shi hensan iinkai (eds), *Tomioka seishijō shi*, jō, 1977, pp. 875–902.
Yomiuri shinbun, 1889.

Appendix

Selected Sources on the Japanese History of Technology (especially on series)

Erich PAUER

1. INTRODUCTION

AS AN APPENDIX to the short introduction at the beginning of this volume, a few selected writings on the history of science and technology, which are important sources for understanding Japan's technological development, are discussed below. Some of them have already been mentioned in the introduction.

2. EARLY BEGINNINGS

A process that we can observe, not only among European countries but also in Japan, is that it was engineers (in most witnessed cases civil engineers) who – well aware of their importance in a newly industrialized country – first perceived themselves as a special (professional) group and became active as such. In Japan, this particularly applied to graduates of the Imperial College of Engineering (Kōbu daigakkō).[1] They already founded their first 'association' in 1879 under the name Kōgakkai (Federation of Engineers) and began to publish their own journal named *Kōgakkai-shi* (Journal of the Federation of Engineers)[2] as early as 1881. The Kōgakkai is regarded as the predecessor of the Doboku gakkai (Federation of Civil Engineers), which was founded in 1914.

1 Founded as Kōgaku-ryō in 1873 and renamed in 1877.

2 The first issues were published under the title Kōgaku sōshi (Library of Engineering).

In the Meiji period, civil engineers were not only numerically the strongest group of engineers, but they were also the ones who very early on took actions against what they perceived as their negative reputation in society. In contrast to this negative attitude, they believed they were rendering important services to society. In order to highlight the importance of engineers and make their merits more widely known, they turned to the history of their profession and began to describe their achievements for the country in a series of historical contributions, written by civil engineers. The series, which has so far received little attention in academic circles, started in 1914 in the first issue of the journal *Kōgaku* (Engineering). The 32 contributions dealt with topics like the construction of the castle in Nagoya, the history of bridge building, the construction of the famous silver pavilion in Kyōto, the casting of the Great Buddha in Nara, the river regulation of Katō Kiyomasa, dam construction, port construction, water supply or construction of fortifications. The vast majority of the contributions focused on Japan; only a few were dedicated to historically significant construction works in other countries.

This was the first series of publications that aimed to draw attention to important historical achievements by civil engineers, but it had little success. The reputation of civil engineers, but also of engineers in general, did not improve through such a series in a specialist journal. In 1918, engineers were obviously still regarded as persons without any insight, understanding or common sense, and were even considered impudent.[3] Similar complaints by engineers about the contempt with which people met their status continued into the 1920s and 1930s.

3. INDUSTRIAL DEVELOPMENT AND HISTORY OF TECHNOLOGY

In the 1920s, especially after the enormous damage and losses caused by the Kantō earthquake of 1923, a movement was set in motion that was dedicated to the history of the individual industries. The public institutions attached particular importance to ensuring that

---

[3] In a significant contemporary criticism, Naoki Rintarō (himself an engineer), in 1918 complained of a special situation, that '[...while] in the eyes of the general society we, the engineers, appear as persons without any insight, understanding or common sense, this is a part of reality; but in this case are we not presented as impudent?' (Naoki Rintarō, *Gijutsu seikatsu yori* (From life with technology), Tōkyō: Shibiru-sha 1918, p. 10).

their history was not forgotten and that their achievements were honoured. Various state-run companies, arsenals, shipyards etc. presented their development in extensive publications in the 1920s. In this context, a series of ten volumes entitled *Meiji kōgyō-shi* (History of industry in the Meiji period) was compiled by the three industrial associations Kōgakkai, Keimeikai and Nihon Kōgakkai and published between 1926–1931.[4] The volumes deal with individual industrial branches and are thus primarily conceived as industrial histories. However, they also contain comprehensive information on the technologies that were imported, developed, introduced and applied in the respective industries. The series is therefore also of great importance for the historian of technology. The following industrial branches are covered: chemical industry, shipbuilding, railroad, construction, electricity, gunpowder, civil engineering, iron and steel, mining, machinery and earth sciences.

As a kind of follow-up publication, the *Taishō kōgyō-shi* (History of industry in the Taishō period), initiated by Japanese and Chinese engineers, was published in three volumes by the associations Kōgakkai and Nihon kōgakkai in Tōkyō in the early years of the Shōwa period.[5]

4. EARLY HISTORY OF TECHNOLOGY IN THE ACADEMIC WORLD – BEFORE WORLD WAR II

In the 1920s, however, an intellectual restlessness was spreading. The crises of the 1920s and, in particular, the world economic crisis, that affected Japan from 1929 to 1932, fuelled the criticism of the economic and social situation. Marxism spread particularly in the milieu of intellectuals, and this was the starting point for a new view on technology in Japan, including the historical perspective. Especially among the philosophers who dealt with materialism and later also with Marxism, there were some who also turned to technology in its material manifestation. One of them was Saigusa Hiroto (1892–1963), who after many years of intensive discussion on the definition and role of 'technology', published *Nihon gijutsu-shi*, the first 'History of technology' in Japan, in 1940.

4 Nihon kōgakkai (ed.), *Meiji kōgyō-shi* (History of industry in the Meiji period), Tōkyō: Meiji kōgyō hakkō-sho, 1926–1931; a reprint by Gakujutsu bunken fukyū-kai, Tōkyō 1968–1972.

5 Nihon kōgakkai (ed.), *Taishō kōgyō-shi* (History of Industry in the Taishō Period), Tōkyō. A new edition by Kōgakkai in 1993–1994 in three volumes was supervised by Haga Namio and published by Hara shobō.

After this outstanding work, he started publishing a first collection of original sources on the history of Japanese science and technology, which, as it is often the case, places a much greater emphasis on Japanese science than on technology.

The series was planned to include 13 volumes (12 volumes plus one index volume) and entitled *Nihon kagaku koten zensho* (Collected classics of Japanese science); a first volume appeared in 1942 – i.e. at a time when the Pacific War was nearing its climax. Due to severe disruptions in the production of books because of the shortage of paper during the war, some volumes were published only after the end of the war in 1948.[6]

While Saigusa focused on the 'classical' sources on the history of science and technology written in the Edo period, a broader approach was taken after the war.

5. HISTORY OF TECHNOLOGY – AFTER WORLD WAR II

Another extensive series on the history of science and technology in Japan, which was still planned during the war, could also, for the most part, only be published in the decades after the end of the war. The Imperial Japan Academy (Teikoku gakushi-in) had already conceived a series entitled *Meiji-zen Nihon kagaku-shi* (History of science in Japan before the Meiji-Restoration) in 1941, but due to the political situation it did not go beyond collections of material at first. The successor of the former academy, called Japan Academy, took up this project again in the post-war period and succeeded in publishing a total of 26 volumes until the 1970s, mainly on the basis of the previously collected material that had been preserved and could be supplemented.[7] Although the focus of most volumes of this series is predominantly laid on the history of science, there are also several volumes on the history of technology, for example, on agricultural technology, fishing techniques, machinery and tools, mining, weapons production, civil engineering and architecture.

A more contemporary view was taken up in the series *Nihon kagaku gijutsu-shi taikei* (Compendium of Japanese history of

---

6 Saigusa Hiroto (ed.), *Nihon kagaku koten zensho* (Collected classics of Japanese science), 10 vols., Tōkyō: Asahi shinbun-sha 1942–1948; reprinted in nine volumes in 1978 by Asahi shinbun-sha.

7 Meiji zen Nihon kagaku-shi kankō-kai (ed.), *Meiji zen …* (var. titles), Tōkyō: Nihon gakujutsu shinkō-kai (var. years).

science and technology) published between 1963 and 1972 in a total of 26 volumes with the focus on the years after the Meiji Restoration (1868).[8] Historians of science, such as Yuasa Mitsutomo (1909–2005), Nakayama Shigeru (1928–2014) and Oka Kunio (1890–1971), as well as historians of technology, such as Yoshida Mitsukuni (1921–1991), Iida Ken'ichi (1926–1997), Yamazaki Toshio (1916–1994) and Hiroshige Tetsu (1928–1975), were involved in this project.

More comprehensive in terms of the number and type of sources is the series *Edo kagaku koten sōsho* (Classical collection of Edo period science)[9] published between 1976 and 1983 and supervised by Aoki Kunio and Iida Ken'ichi, two well known historians of technology. It presents numerous sources from the Edo period, covering several branches of industry such as mining, agriculture, textiles, electricity, carpentry, silkworm breeding and coin production. It thereby includes many extremely valuable and rare sources, which are hitherto nearly unknown and usually cannot be found in archives or libraries. In contrast to other series, also mentioned here, this one clearly focuses on the history of technology.

In the 1980s, several, albeit less comprehensive, series on the history of technology appeared, which also bear this term in their titles. These include, for example:

Kōza Nihon gijutsu no shakai-shi (A social history of Japanese technology),[10] published 1983–1986. The volumes of this series are divided into the following categories: agriculture, salt and fishing, textiles, ceramics, mining and metallurgy, civil engineering, architecture, transport and traffic, and biographies.

Gijutsu no shakai-shi (A social history of technology),[11] published by Yūhikaku 1982–1990. The topics dealt with by different authors range from technology and society in ancient and medieval times, to the early modern period and indigenous technology, the Meiji period and technology transfer, development and heavy

---

[8] Nihon kagaku-shi gakkai (The History of the Science Society of Japan) (ed.), *Nihon kagaku gijutsu-shi taikei* (Compendium of Japanese history of science and technology), 26 vols., Tōkyō: Dai-ichi hōki shuppan, 1963–1972.

[9] Aoki Kunio, Iida Ken'ichi et al. (eds and comp.), *Edo kagaku koten sōsho* (Classical collection of Edo period science), 46 vols., Tōkyō: Kowa shuppan, 1976–1983.

[10] Nagahara Keiji & Yamaguchi Keiji (repres. eds), *Kōza Nihon gijutsu no shakai-shi* (A social history of Japanese technology), 8 + 2 vols., Tōkyō: Nihon hyōron-sha, 1983–1986.

[11] *Gijutsu no shakai-shi* (A social history of technology), 6 + 1 vols., Tōkyō: Yūhikaku, 1982–1990.

industry and its contradictions, industrial society and technology, all the way to present society and technological innovation, concluded with a 'data guide'.

Bijuaru-ban: Nihon no gijutsu hyakunen (A century of technology in Japan, illustrated),[12] published 1988–1989. In this series the individual industrial sectors are presented visually: mining and energy, steel and metal processing, shipbuilding and railway, aircraft and automobiles, telecommunication, radio and TV, architecture and civil engineering, and machinery and electronics.

Nihon no gijutsu (Series: historical technology of Japan),[13] published 1988–1991. In these volumes, individual authors discussed specific industrial sectors or products: industrial technology, Yawata-steel works, Japanese sake, Chinese medicine, rotary engines, furniture, woodworking tools, the Ōsaka arsenal, Japanese films, beer and motorcycles.

As a sequel to *Nihon kagaku gijutsu-shi taikei* (see above), a series entitled *Tsū-shi Nihon no kagaku gijutsu* (engl. title: The History of Science and Technology in Contemporary Japan),[14] was published by Nakayama Shigeru, Gotō Kunio and Yoshioka Hitoshi 1995–1999. The volumes cover the period from 1945 to 1995. The series is divided into: Occupation period (1945–52), independent period (1952–59), high-growth period (1960–69), transformation period (1970–79) and, in two volumes (5–1 and 5–2), the international period (1980–95). The supplementary volume contains the index and chronological tables.

A follow-up series titled *Shin tsūshi – Nihon no kagaku gijutsu* (A social history of science and technology in Japan at the turn of the century),[15] spanning the years from 1995 to 2011, was published under the representative editor Yoshioka Hitoshi in 2011–2012.

---

12 *Bijuaru-ban: Nihon no gijutsu 100nen* (A century of technology in Japan, illustrated), 7 vols., Tōkyō: Chikuma shobō 1988–1989

13 *Nihon no gijutsu* (Series: historical technology of Japan), 11 vols., Tōkyō: Dai-ichi hōki shuppan, 1988–1991.

14 Nakayama Shigeru (repres. ed.), *Tsū-shi Nihon no kagaku gijutsu* (Engl. title: The Social History of Science and Technology in Contemporary Japan), 5 + 1 vols., Tōkyō: Gakuyō shobō 1995–1999. An English edition under the same English title, comprising 4 volumes, was published by Trans-Pacific PR 2001–2006. The books cover the years from 1945 to 1979.

15 Yoshioka Hitoshi (repres. ed.), *Shin tsūshi Nihon no kagaku gijutsu* (A Social History of Science and Technology in Japan at the Turn of the Century), 5 vols., Tōkyō: Hara shobō 2011–2012.

List of Contributors

HASHIMOTO Takehiko is Professor of History of Science and Technology at the University of Tōkyō. He has studied the history of science and technology both as an undergraduate and graduate student at the University of Tokyo. He received his Ph.D. from Johns Hopkins University in 1991 for his research on the early history of aerodynamics and aeronautics. Since then, he has been a faculty member of the University of Tōkyō teaching the modern history of science and technology. He has widely published on the history of modern science and technology, as well as on the origin of punctuality in modern Japan and the history of the standardization in technological systems. His research interests include the history of physical sciences and technology in the twentieth century and the history of the establishment and evolution of safety standards.

Annick HORIUCHI is professor at Paris Diderot University, and member of the Research Centre of East Asia Civilizations (CRCAO). Professor Horiuchi specializes in the intellectual history and history of science of early modern Japan. She is particularly interested in the contacts between Chinese, Western and Japanese ways of thinking in eighteenth century Japan. She is the author of *Japanese Mathematics in the Edo Period (1600-1868): A study of the works of Seki Takakazu (?-1708) and Takebe Katahiro (1664-1739)* (Science Networks. Historical Studies), Birkhäuser Basel, 2010, and has co-edited with Matthias Hayek: *Listen, Copy, Read: Popular Learning in Early Modern Japan*, Japanese Studies Library, Brill, 2014.

ITŌ Mamiko received her Ph.D. from Gakushūin University, Tōkyō, for her dissertation on research concerning the Meiji Government and the World Exposition. She is currently Professor of Modern Japanese History at the International Centre of Gakushūin University. In her research she focuses mainly on expositions, world expositions and museums.

Regine MATHIAS studied Japanese History and History at Ruhr-University Bochum and Kyūshū University and obtained her Ph.D. from the University of Vienna with a thesis on the development of wage labour in Japanese coal mines. She taught at several universities and worked as a professor of Japanese history at Ruhr-University Bochum from 1996 to 2016. Since her retirement, she has been working at the Centre Européen d'Études Japonaises d'Alsace (CEEJA). Her main field of research is Japanese social and economic history, with a focus on Japanese mining and labour history. She has published on labour in Japanese coal-mines, Japanese mining picture scrolls and their value as historical sources as well as on German-Japanese relations and gendered working patterns in pre-war Japan.

NISHIYAMA Takahiro studied sociology, ethnic studies and intercultural didactics, with foci on industrial and cultural sociology, at the University of Göttingen, Germany. He received his Ph.D. in Japanese Studies from the University of Marburg, Germany, in 2009 with his dissertation on "Technical Human Capital Formation in the Yokosuka Shipyard at the Beginning of the Industrialization of Japan". It dealt with the technology transfer and the transfer of the vocational education and training system from the West to Japan from the end of the Tokugawa era to the end of the nineteenth century. He was a research fellow at the University of Marburg, Germany, from 2006 to 2008, and lecturer at the University of Bonn, Germany from 2008 to 2018. His main research areas are the social and industrial histories of modern Japan and Germany, and social and cultural studies of contemporary Japan and Germany.

Erich PAUER is Professor Emeritus at the University of Marburg, Germany. Born in 1943 in Vienna, he obtained a technical education, before majoring in Japanese studies at the University of Vienna, where he received his Ph.D. in 1972. After his post-doctoral research at the University of Tōkyō, he took up a lectureship at the University of Vienna in 1974 and moved to the University of Bonn, Germany, in 1977. He became full professor at the University of Marburg in 1987, where he established the Centre for Japanese Studies. Since his retirement from this university in 2008, he has been working at the Centre Européen d'Études Japonaises d'Alsace (CEEJA). His main interests include the history of technology in Japan in the early-modern period and the period of Japan's industrialization, and Japanese economic history. He has also published on German-Japanese relations in pre-war Japan.

SUZUKI Jun is currently a professor at the Graduate School of Humanities and Sociology and the Faculty of Letters at the University of Tōkyō. He has studied history at the University of Tōkyō and earned his Ph.D. from this university in 1995 with a study on the "Machine industry in the Meiji era". Later, he focused on the industrial policy in the Meiji period and published on the Ministry of Public Works (*Kōbushō*), on engineers and on science and technology policy in modern times.

Index

Note: Page numbers followed by "*n*" refer to notes, "*f*" denotes to figures and "*t*" denotes to tables.

For Product Safety Concerns and Information please contact our EU representative GPSR@taylorandfrancis.com Taylor & Francis Verlag GmbH, Kaufingerstraße 24, 80331 München, Germany

Batch number: 10722102

Printed by Printforce, the Netherlands